International Anesthesiology Clinics

International Anesthesiology Clinics (ISSN 0020-5907). Published quarterly by Little, Brown and Company, 34 Beacon Street, Boston, Massachusetts 02108-1493. Send address changes and subscription orders to Little, Brown and Company, Subscription Department, PO Box 2033, Langhorne, PA 19047-9480; (800) 628-4221. Subscription rates per year: personal subscription, U.S. and possessions, $81; foreign (includes Mexico), $105; Canada, $92; institutional, U.S., $99; foreign, $129; Canada, $109. Special rates for students, interns, and residents per year: U.S., $55; foreign, $75; Canada, $65. Single copies: $30 for subscribers, $38 for nonsubscribers. In Japan please contact our exclusive agent: Medsi, 1-2-3 Yushima, Bunkyo-ku, Tokyo 113, Japan. Subscription rates per year in Japan: individual, ¥22,200; institutional, ¥26,200 (air cargo service only). Second-class postage paid at Boston, Massachusetts, and at additional mailing offices.

Postmaster: Send address changes to International Anesthesiology Clinics, PO Box 2033, Langhorne, PA 19047-9480.

International Anesthesiology Clinics is indexed in *Index Medicus*, *Current Contents/Clinical Practice*, *Excerpta Medica*, and *Current Awareness in Biological Sciences*.

International Anesthesiology Clinics

Recovery from Anesthesia

Volume 29
Number 2
Spring 1991

Little, Brown and Company
BOSTON

Editor-in-Chief

Philip W. Lebowitz, MD
Long Island Jewish Medical Center
Albert Einstein College of Medicine

Editorial Office

Department of Anesthesiology
Long Island Jewish Medical Center
270-05 76th Avenue
New Hyde Park, NY 11042

Publisher

Little, Brown and Company,
Boston, Massachusetts

Publishing Staff

Publisher
Lynne Herndon

Executive Editor
Christine F. Lamb

Managing Editor
Sheila Gambill

Production Manager
Fredda Purgalin

Sales and Marketing Manager
Anne Orens

Contents

Contributing Authors

Paul H. Alfillé, MD
Department of Anesthesia
Massachusetts General Hospital
Boston, MA 02114

Richard Austin, MD
Department of Anesthesiology
Long Island Jewish Medical Center
270-05 76th Ave
New Hyde Park, NY 11042

Daniel Curtis, DO
Department of Anesthesiology
Oregon Health Sciences University
Portland, OR 97201

Eran Geller, MD
Department of Anesthesiology and Intensive Care
Tel Aviv Sourasky Medical Center
Ichilov Hospital
6 Weitzman St
Tel Aviv 64239
Israel

Pinchas Halpern, MD
Department of Anesthesiology and Intensive Care
Tel Aviv Sourasky Medical Center
Tel Aviv University Sackler Faculty of Medicine
Tel Aviv
Israel

Roger Mecca, MD
Department of Anesthesiology
Danbury Hospital
Hospital Ave
Danbury, CT 06810

Ramesh Patel, MD
Department of Anesthesiology
Children's National Medical Center
111 Michigan Ave, NW
Washington, DC 20010

Shubhinder Puri, MD
Department of Anesthesiology
Yale University School of Medicine
New Haven, CT
Norwalk Hospital
Norwalk, CT 06856

Frederic M. Ramsey, MD
Department of Anesthesiology
Winchester Hospital
Winchester, MA 01890

Linda Jo Rice, MD
Department of Anesthesiology
Children's National Medical Center
111 Michigan Ave, NW
Washington, DC 20010

Carl Rosow, MD, PhD
Department of Anesthesia
Massachusetts General Hospital
Boston, MA 02114

Lalage D. Sanders, PhD
University of Wales
College of Medicine
Heath Park
Cardiff CF4 4XW
Wales

Aryeh Shander, MD
Department of Anesthesiology
Albert Einstein College of Medicine
Bronx, NY
Norwalk Hospital
Norwalk, CT 06856

Wendell C. Stevens, MD
Department of Anesthesiology
Oregon Health Sciences University
Portland, OR 97201

Preface

Although functionally a continuum, an anesthetic may be conceptually considered as consisting of separate phases of induction, maintenance, and emergence. None of these phases has been demonstrated to be intrinsically safer or less problematic than any other, yet both the lore and literature of anesthesiology have disproportionately emphasized the process and pitfalls of induction. Maintenance has largely been ignored, and emergence is commonly welcomed as the end of the vigil, characterized by the elimination of confounding and potentially deleterious physiological and pharmacological influences—an anesthetic "Miller time."

Our paradigmatic vigilance relaxes. We may, in uncomplicated cases, disconnect the electrocardiogram, blood pressure cuff and pulse oximeter as the patient resumes spontaneous ventilation just prior to tracheal extubation. We lower our guard. We start thinking about lighter concerns or the next case as we wheel the patient, unmonitored and breathing room air, to the recovery area, where monitoring will be resumed. And why not? Our skills well honed and our judgment clear, we know which patients are likely to get into trouble, and this isn't one of them. Or is it? Is routine emergence from anesthesia routine—and uncomplicated?

This issue of *International Anesthesiology Clinics* addresses these questions but does not presume to answer them definitively. Although a body of knowledge dealing with the pharmacology of antagonists to benzodiazepines, neuromuscular blocking drugs, and opioids is accumulating, information about the physiological consequences of emergence from the anesthetized state is sparse. And similarly, whereas description and measurement of postoperative psychological effects exist, better understanding of cognitive function on return from this neurological trespass known as anesthesia requires extensive further investigation.

Anesthesiologists are now being confronted with a concept of emergence as a destabilizing influence on homeostasis. We are increasingly working to smooth the rough edges and eliminate emergence stress. It seems reasonable to conclude that better use of even our current monitoring capacity affords us the opportunity to minimize risk from emergence effects, which, unwitnessed and unappreciated, might lead to patient harm.

P.W.L.

Recovery from General Anesthesia

Daniel Curtis, DO
Wendell C. Stevens, MD

The immediate goal of anesthetic recovery is return of the patient's ability to independently maintain respiratory and circulatory functions. The eventual return of all mental, sensory, and motor abilities to preanesthetic levels is ordinarily an achievable goal following inhalation anesthesia. Complete recovery, signified by elimination of all anesthetic molecules from the body, eventually takes place, but the time at which this occurs bears little relation to clinical signs observed during recovery.

Removal of all anesthetic molecules after inhalation anesthesia requires several days [1]. The amount of drug to be eliminated depends mainly on the dose and duration of anesthetic delivery as well as the solubility of the drug used for anesthesia [2]. Most of the anesthetic is removed from the body in the first few minutes after discontinuation of drug delivery with agents having poor solubility [3]. The reservoir of these poorly soluble drugs in blood and tissues is small, and lung washout rapidly decreases their alveolar concentration, which in turn favors prompt elimination from tissues. As anesthetic solubility increases, the reservoir of agent in tissues also increases. Removal of drug from tissues continues to replenish alveolar anesthetic. With highly soluble drugs, elimination may require hours or even days. The anesthetic levels for a large part of the elimination period are so low that the patient is unaware of the anesthetic, and few if any physiological effects of the drug can be accurately measured. Patients may have symptoms such as malaise or inability to concentrate following anesthesia, but they are not necessarily due to residual anesthetic agent remaining in tissues [4].

There is only one clearly defined point in recovery that relates alveolar concentration to an easily reproducible variable: MAC-Awake [5]. The alveolar concentration at which 50% of patients respond to a simple command such as "open your eyes" defines MAC-Awake. MAC-Awake is approximately one-half the minimum alveolar concentration (MAC) determined by response to a surgical incision. The MAC-Awake point is passed during emergence from every general anesthetic even though it is not routinely determined precisely. It cannot be determined precisely during routine awakening because the brain anesthetic concentration lags behind the rapidly decreasing alveolar concentration during recovery. Determination of MAC-Awake would require stepwise decreases of alveolar concentration with sufficient time at each step to allow brain and alveolar anesthetic levels to equilibrate.

MAC-Awake is important in the context of recovery because a patient at this dose can maintain a natural airway without assistance. Patients should likewise be able to respond to commands to breathe deeply unless some other impediment to breathing exists. Hypoxic ventilatory response may still be depressed at sub-MAC-Awake levels, but the carbon dioxide ventilatory response should be near normal [6].

The circulatory effects of anesthetizing doses of these drugs are well known, but effects of MAC-Awake or lower levels are virtually unstudied. One might postulate that patients who are hypovolemic, have inadequate circulatory baroreceptor function, or have myocardial insufficiency will respond adversely to even subanesthetic doses of anesthetic drugs. However, supporting data relating early postoperative circulatory instability to alveolar anesthetic concentration are unavailable.

Mental, sensory, and motor functions return between the point at which recovery is sufficient for the patient to maintain an airway without assistance and the time when all the anesthetic is eliminated. Mental and psychological effects possibly related to anesthesia may persist for several days following anesthesia even if surgery is not done [4]. It is evident by the onset of pain after surgery that sensation returns on awakening. Unfortunately, the more promptly recovery occurs, the sooner pain begins. This situation has led to discussion of whether it is better for a patient to awaken quickly and require analgesic drugs soon after surgery or to receive narcotics intraoperatively and awaken more slowly and comfortably. The modern inhalation agents do not have important, persistent analgesic properties at subanesthetic concentrations. The current preference is for anesthetics of low solubility because they enable prompt recovery and have putatively lower liver and kidney toxicity; early onset of pain is a price patients pay for our use of these drugs. Muscle function, apart from disturbances of coordination or equilibrium, returns to awake levels by the time the patient reaches MAC-Awake.

■ The Pharmacokinetics of Recovery

A major attribute and advantage of inhalation agents is their entry into and removal from the body primarily by way of the lungs. A fraction of inhaled anesthetics is metabolized, but this plays little role in the return of consciousness when anesthetic delivery is discontinued [2]. Anesthetic metabolism, however, will be considered as a factor affecting decrease of alveolar concentration late in recovery. The small amount of anesthetic that is eliminated through the skin is inconsequential in the context of recovery [7].

Because anesthetic entry and removal occur by way of the lungs, it is appropriate to discuss the factors governing alveolar concentration during recovery. We can now routinely and accurately measure alveolar concentrations as a guide to recovery. An important caveat exists when measuring end-tidal concentration during awakening: tissue concentrations are not as low as the alveolar concentration during recovery. One is likely to overestimate the extent of recovery based on analysis of alveolar concentrations alone. Eger [2] and, more recently, Dale and Brown [8] have analyzed in detail the physiological and pharmacological factors governing recovery.

Inspired Anesthetic Concentration

For recovery to be most rapid, the inspired anesthetic concentration should be zero. Two reasons this ideal may not exist include rebreathing of exhaled gases and release of gases dissolved in the components of the anesthetic circuit into the respired gases. Rebreathing is prevented by providing inflow rates equal to or in excess of minute volume when using a circle system, or inflow rates 2½ times minute volume with a T-piece system—e.g., the Bain circuit. With the circle system, the circuit should first be emptied of anesthetic gases to reduce the time required to rid the circuit of high concentrations of agent. This maneuver is unnecessary with a T-piece system because change in inspired concentration is achieved instantaneously with change in delivered gas concentrations. Only small amounts of anesthetic are dissolved in the plastic components of the circuit—far less than in the rubber components that used to be employed [9]. Newer, less soluble agents dissolve in these components to a smaller extent than did highly soluble anesthetics. Soda lime also absorbs anesthetics, especially when the granules are dry [10]. Therefore, the absorbent is a source of anesthetic. Such release delays recovery only slightly and not perceptibly.

Alveolar Ventilation

Gases are removed by ventilation, and an increase in ventilation hastens recovery. With all agents, ventilation initially lowers the alveolar anesthetic concentration dramatically because the effect of ventilation is unopposed by delivery of gases from the tissues to the lung. After the first few minutes, the effect that changes in ventilation have on alveolar concentration is complex and exhibits a logical and interesting relationship to solubility and to duration of anesthesia [3]. As anesthetic solubility and duration of anesthesia increase, the impact of changes in ventilation becomes greater late in the recovery period. Thus, augmenting breathing assists elimination of poorly soluble drugs to the greatest extent in the first few minutes of recovery regardless of the duration of anesthesia because the amount of drugs in tissues is small even after prolonged anesthesia. At the other extreme, hyperventilation in the first few minutes of recovery does little to speed elimination of a highly soluble drug if the duration of anesthesia has been long, because the amount of drug in tissues is so great. When both anesthetic solubility and duration of anesthesia are intermediate between these extremes, the time when alteration of breathing affects recovery is also intermediate; that is, in the range of 10 to 30 minutes after discontinuing anesthesia (Fig 1).

In practice, ventilation may vary during recovery. Various combinations of depression from inhaled anesthetic, stimulation of breathing by pain, and depression from analgesic drugs can alter the normal, gradual decrease in alveolar anesthetic (Fig 2).

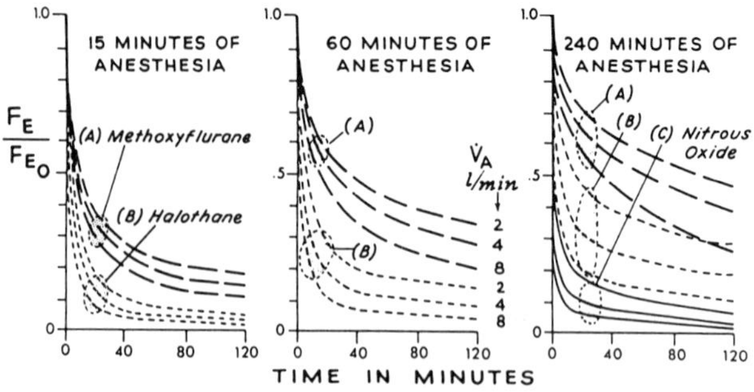

Fig 1. *Simulated F_E/F_{EO} curves, showing the effects on recovery of alveolar ventilation (2, 4, and 8 L/min) after 15, 60, and 240 minutes of ancsthesia. F_E/F_{EO} is the relation of alveolar gas tension at each time in recovery divided by the alveolar gas tension at the start of recovery. (Reproduced with permission from Stoelting RK et al [3].)*

Diffusion Hypoxia

When delivery of anesthetic gases containing nitrous oxide ceases, dilution of alveolar gases by nitrous oxide occurs because it is delivered from blood to alveoli more rapidly than nitrogen is taken up in the reverse direction. Hypoxemia can occur due to dilution of alveolar oxygen [11]. In the context of rate of recovery, dilution of alveolar carbon dioxide is important because decreased alveolar ventilation delays recovery from anesthesia. Hypoxia is prevented by use of high inspired oxygen concentrations; hypoventilation is prevented by assisted breathing.

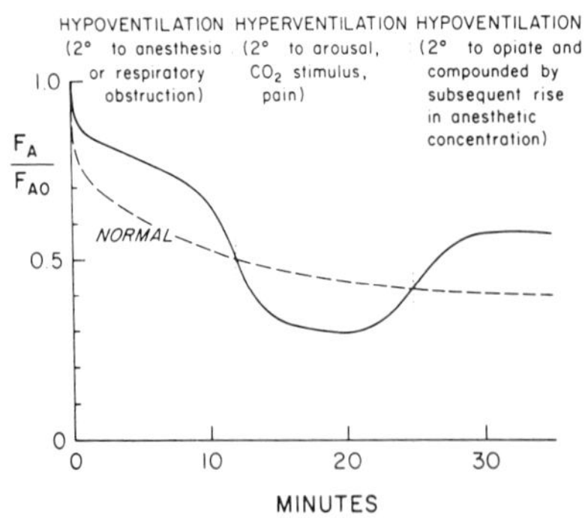

Fig 2. *Ventilation may vary during the course of recovery, causing deviations* (solid line) *from an idealized* (dashed line) *representation of the decline in alveolar anesthetic concentration. F_A/F_{AO} is the ratio of alveolar partial pressure at a given time to that immediately before discontinuation of anesthetic administration. Hypoventilation during the first 12 minutes might be caused by anesthetic depression or obstruction, or by a "reset" of the CO_2 sensor that would produce a retarded rate of fall. In the next 12 minutes hyperventilation secondary to awakening, an increased perception of elevated CO_2, and/or pain might accelerate the rate of fall. Subsequently, administration of a narcotic might relieve pain. Both relief of pain and the opioid's direct depressant effect might reduce ventilation and allow the alveolar anesthetic concentration to rise again. (Reproduced with permission from Eger [2]; © Edmond I. Eger II, MD.)*

Posthyperventilation Depression of Breathing

The ventilatory response to carbon dioxide and the apneic CO_2 threshold are depressed by inhaled agents. Therefore, the arterial Pco_2 must be increased above normal during anesthesia if spontaneous breathing is to occur. If the patient has been hyperventilated for a long period so that carbon dioxide stores are decreased, some time is required for CO_2 stores to be replenished and for arterial Pco_2 to rise to ventilatory threshold levels [12]. The resultant hypoventilation delays awakening. This circumstance can be avoided in at least two ways. One method is to maintain arterial Pco_2 near the apneic point throughout anesthesia. Another method is to intentionally decrease alveolar ventilation late in the anesthetic period to restore arterial Pco_2 nearer to awake levels.

Cardiac Output and Tissue Perfusion

If cardiac output decreases during recovery without a corresponding decrease in alveolar ventilation, alveolar anesthetic concentration falls more rapidly than otherwise because less anesthetic is delivered to the lung [2]. However, awakening may not seem to be accelerated because reduced tissue blood flow slows removal of agent from the tissues. If cardiac output increases, alveolar anesthetic levels are sustained unless alveolar ventilation increases. At the same time, increased tissue perfusion favors removal of anesthetic from tissues.

In practice it is uncommon for one factor to change independent of changes in other factors. For example, a patient in shock who exhibits hyperventilation and low cardiac output but near-normal cerebral perfusion should awaken quickly when anesthetic delivery stops. A patient who is hyperthermic, with increased ventilation and increased cardiac output that is distributed proportionately to all tissues, may awaken at about the same rate as a normothermic patient. Reduction of cerebral blood flow by hyperventilation is an example of decreased tissue perfusion delaying recovery; removal of even poorly soluble agents from the brain will be delayed in this instance.

Ventilation/Perfusion Abnormalities

Shunts of blood in the lung delay elimination of the anesthetic. The effect is greatest for agents of low solubility. This factor does not appear to be of great importance to the rate of recovery.

Anesthetic Solubility

The rate of induction and emergence from anesthesia is inversely related to anesthetic solubility. It is the low solubility of the newest anesthetic drugs, desflurane and sevoflurane, that captured the attention of investigators [13, 14]. Alveolar concentration decreases rapidly, and awakening occurs promptly because tissue anesthetic levels fall so quickly with the new anesthetics.

Hypothermia is a common event that increases anesthetic solubility and can also delay emergence in other ways. Concomitantly, ventilation, cardiac output, and tissue perfusion decrease. Each of these decreases the rate of recovery on a pharmacokinetic basis. Awakening may also be delayed because the anesthetic requirement is reduced [15]. A decrease in MAC-Awake during hypothermia means the alveolar concentration must decrease below expected normothermic levels before the patient will respond.

Anesthetic Metabolism

Metabolism is a major factor in determining the rate of decrease of alveolar concentration of some anesthetics, although clinical awakening is not accelerated by anesthetic metabolism. Metabolism continues throughout the course of anesthesia, but the fraction metabolized is very small at anesthetizing concentrations. Metabolism becomes a major factor in halothane elimination when the halothane alveolar concentration falls below about 1/30 MAC [16]. The combination of moderate solubility and susceptibility to metabolism result in metabolism of as much as 15 to 20% of the amount of halothane administered in a 1- to 2-hour anesthetic. Moderate solubility leads to persistence of low tissue halothane levels for long periods of time. Although these features do not affect clinical awakening, they do account for the greater rate of decline of alveolar halothane concentration compared with that of the less soluble isoflurane as recovery continues into the third and fourth hours (Fig 3) [1]. Metabolism has less affect on recovery from the newer, less-soluble drugs, desflurane and sevoflurane, because these agents are eliminated from the body so quickly.

Duration of Anesthesia

It is intuitively obvious that as the duration of anesthesia increases, the time required for recovery also increases [17]. This relationship pertains because prolonged anesthesia allows anesthesia levels in all tissues to more closely approximate the alveolar concentration. In this situation, when anesthetic deliv-

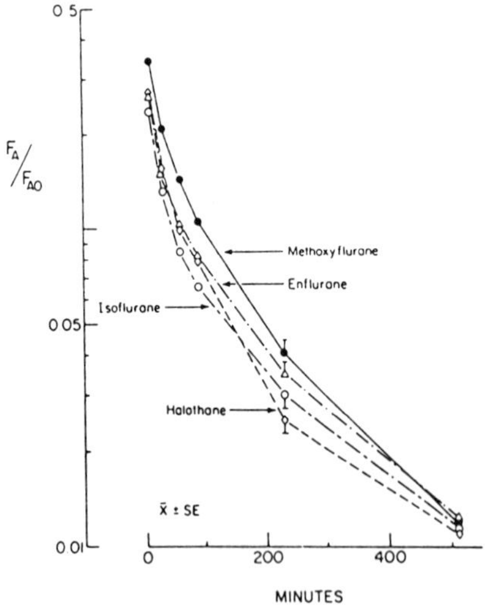

Fig 3. *The decline in alveolar anesthetic concentration (F_A/F_{AO}) is plotted against time (means for all data ± SE for 2- and 4-day values). Note that F_A/F_{AO} is plotted on a logarithmic scale. (Reprinted with permission from the Anesthesia Research Society from Carpenter et al [1].)*

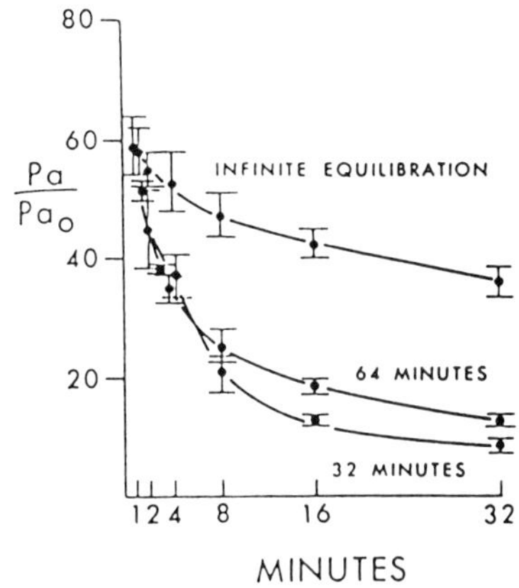

Fig 4. *Isoflurane arterial washout curves for three groups of subjects. (Infinite equilibration—5 subjects anesthetized with isoflurane and N_2O for 6 to 7 hours. Sixty-four minutes—4 subjects exposed to a subanesthetic isoflurane concentration for 64 minutes. Thirty-two minutes—3 subjects exposed to a subanesthetic isoflurane concentration for 32 minutes. Increasing the time of exposure to isoflurane results in increasing tissue saturation and a decreased rate of excretion. (Reproduced with permission from Cromwell et al [17].)*

ery is discontinued, venous and alveolar anesthetic levels tend to be maintained from tissue stores (Fig 4). Recent laboratory studies have aided our understanding of how alveolar concentration, anesthetic solubility, and duration of anesthesia interact to affect recovery [18]. Increasing the duration of anesthesia prolongs recovery time for all anesthetics. This relationship is especially strong for agents of intermediate (halothane) or high (methoxyflurane) solubility. Recovery times from anesthesia with desflurane, nitrous oxide, and sevoflurane are influenced minimally by duration or alveolar concentration of the anesthetic (Fig 5). Recovery from isoflurane lies intermediate between the extremes described for the other agents. These data mean that continuation of desflurane and sevoflurane at anesthetizing doses to the completion even of long operations, doses great enough to produce some degree of systemic arterial hypotension, should not unduly prolong recovery beyond the 2- to 5-minute recovery times noted after short procedures. By contrast, continuation of halothane anesthesia to the end of a 2- to 3-hour operation results in a recovery time of approximately 10 minutes before the patient responds to simple commands (Fig 6) [17]. A longer recovery time ensues if anesthesia time is longer.

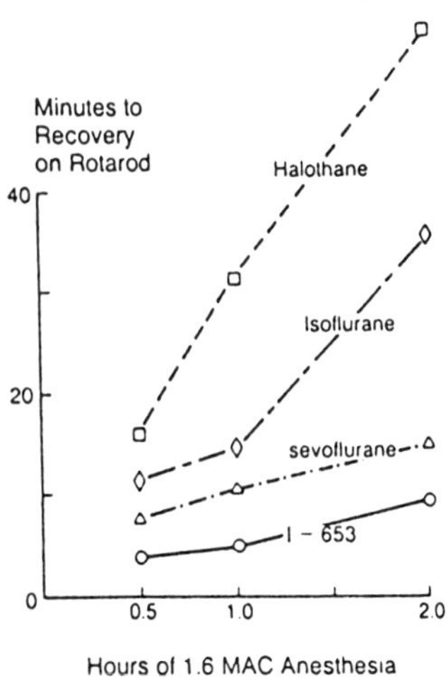

Fig 5. *An increase in the duration of anesthesia (dose constant at 1.6 minimum alveolar concentration) increased the minutes following anesthesia before the rats could maintain balance on a rotating rod. The time to recovery was a function of the solubility of the anesthetic. The slope of the increase was proportional to the solubility of the anesthetic, being steepest with the most soluble agent, halothane, and least so with the least soluble agent, I-653. (Reprinted with permission from the Anesthesia Research Society from Eger et al [18].)*

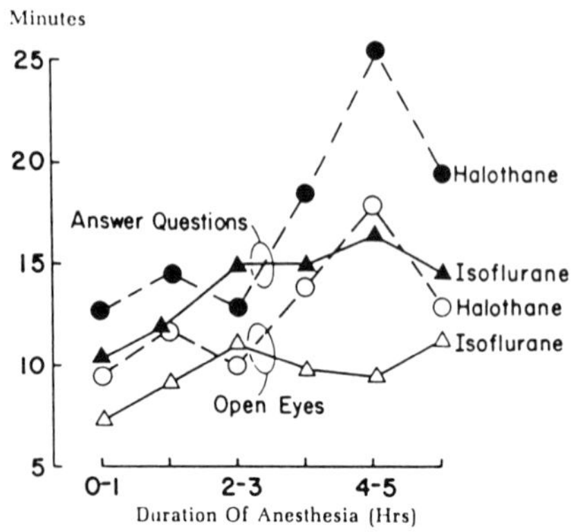

Fig 6. *Relationship of time of appearance of signs of recovery in patients anesthetized with halothane* (open and closed circles) *or isoflurane* (open and closed triangles). *Approximately 900 patients were anesthetized for various lengths of time with isoflurane, and a similar number with halothane. Premedication, anesthetic concentration, and the use of supplementary agents were selected at the discretion of the anesthesiologist in charge of the case. The times from discontinuation of the anesthetic to opening of the eyes and to answering questions were measured. (Adapted from the Isoflurane New Drug Application, pp. 626–627.) (Reproduced with permission from Eger et al [19].)*

■ Clinical Decisions and Problems Accompanying Recovery from General Anesthesia

Combinations of Agents

We have described some features of recovery from inhalation anesthesia, idealized because recovery from a single agent conforms to pharmacokinetic principles in a straightforward manner. In practice, various combinations of drugs are generally used, making it nearly impossible to describe all the patterns of recovery that occur. Recovery patterns assessed by time to awakening or time to hospital discharge, or by occurrence of nausea, need for pain medication, or other variables have been compared in many regimens [18–20]. Almost any regimen can be managed safely [21].

Some strengths and weaknesses of certain methods of anesthesia with respect to subsequent recovery patterns have emerged. For example, the use of some narcotic agent appears to improve immediate postoperative comfort [22]. On the other hand, drugs with sedative but not analgesic properties may lead to greater anxiety and "dis-ease" on recovery. Also, techniques that rely heavily on inhalation agents are associated with a high incidence of nonthermogenic tremors during emergence, although any technique can be associated with this phenomenon [23].

Timing of Extubation

Timing of tracheal extubation, perhaps more than any other aspect of emergence, is dependent on the wisdom gained from clinical experience for its successful management. Extubation at deep levels of anesthesia prevents adrenergic and airway responses to tracheal and laryngeal stimulation but leaves the airway unprotected. Extubation at intermediate levels of anesthesia is to be avoided because airway reflexes may be heightened and sustained laryngospasm may occur immediately after extubation. In this setting, even though laryngospasm occurs, chemical control of breathing may still be depressed, so that little inspiratory effort is made. Relief of the laryngospasm may occur only after the patient has become hypoxic or after a neuromuscular blocking drug or intravenous anesthetic has been administered. Extubation at a light level of anesthesia, so-called awake extubation, offers the advantage of recovered airway reflexes to prevent pulmonary aspiration and assures return of near-normal breathing and circulatory function. However, the presence of the endotracheal tube as the patient awakens may lead to an intense adrenergic response, outpouring of upper airway secretions, repetitive coughing, and sustained straining or Valsalva maneuvers. The latter can be detrimental to the integrity of the operative site, leading to wound disruption or bleeding.

Depression of airway reflex responses and some attenuation of adrenergic responses can be obtained with administration of parenteral narcotics or local anesthetics. If the adrenergic response is clearly detrimental to a patient, alpha- and/or beta-blocking drugs or other vasodilators can be used prior to or during recovery.

Positioning during Recovery

Airway and circulatory management dictate positioning during recovery in the absence of specific postsurgical positioning requirements. If the patient cannot maintain an airway without assistance when in the supine position, the patient should be placed in a lateral decubitus position. A patient who is extubated but has not reached the MAC-Awake level should be recovered in the lateral or prone position with the head dependent in relation to the chest to reduce the risk of pulmonary aspiration. With regard to circulatory adequacy, the risk of persistently depressed baroreceptor function dictates maintenance of the patient in a horizontal position, at times with the legs slightly elevated until transfer to the Post Anesthesia Care Unit (PACU) is complete and circulatory stability is assured.

The Patient Who Is Slow To Awaken

Patients occasionally remain anesthetized well beyond the expected recovery time; often the reason for this is not clear in individual cases. Relative overdose of anesthetic agents may not be apparent, and sufficient time may have elapsed for normal elimination or redistribution of drugs. The following questions should be asked in the situation of persistent somnolence:

1. Has ventilation been sufficient to eliminate the inhaled agents and to prevent severe hypercarbia? Has the patient been hyperventilated for a long period, resulting in hypocarbia?
2. Is oxygenation impaired?
3. Is the patient severely acidotic or alkalotic?
4. Has cardiac output (and perfusion) been sufficient to deliver drugs from tissues to the lung or liver?
5. Is the patient hypothermic?
6. Was the patient receiving medication prior to anesthesia that could have decreased anesthetic requirement, e.g., lithium or clonidine?
7. Was premedication given that has prolonged central effects, e.g., anticholinergic, benzodiazepine, or butyrophenone?

8. Was a combination of intravenous and inhaled agents given, leading to additive or synergistic effects?
9. Was the neuromuscular blocking drug adequately reversed?
10. Has advanced age contributed to the effect of the administered drugs?
11. Is the patient diabetic and receiving insulin? Is the patient hyperglycemic or hypoglycemic?
12. Is the patient hyponatremic, hypokalemic, hyperkalemic, hypomagnesemic, hypermagnesemic, or hypophosphatemic?
13. Has an intercurrent central nervous system event occurred, e.g., a stroke?

■ Summary

The pattern of recovery from anesthesia is ordinarily quite predictable. Using a limited number of drugs to provide anesthesia further improves predictability; it is more difficult to assign a cause for delayed emergence when multiple agents have been administered.

We believe the recovery period is often more difficult to manage safely than induction and maintenance of anesthesia. During this phase, the effects of surgical trauma are superimposed on the patient's preexisting disease state, the anesthetic level is changing rapidly, respiratory function has not returned to normal, and circulatory reflexes may not have been regained. Because the patient is being moved from the operating room and care of the patient is being transferred from one person or team to others, vigilance and treatment may also be impaired. It is a period of care that warrants further study.

■ References

1. Carpenter RL, Eger E I II, Johnson BH, et al. Pharmacokinetics of inhaled anesthetics in humans: measurements during and after the simultaneous administration of enflurane, halothane, isoflurane, methoxyflurane, and nitrous oxide. Anesth Analg 1986;65:575–582
2. Eger E I II. Recovery from anesthesia. In: Eger E I II. Anesthetic uptake and action. Baltimore: Williams & Wilkins, 1974:228–248
3. Stoelting RK, Eger E I II. The effects of ventilation and anesthetic solubility on recovery from anesthesia: an in vivo and analog analysis before and after equilibrium. Anesthesiology 1969;30:290–296
4. Davison LA, Steinhelber JC, Eger E I II, Stevens WC. Psychological effects of halothane and isoflurane anesthesia. Anesthesiology 1975;43:313–324
5. Stoelting RK, Longnecker DE, Eger E I II. Minimum alveolar concentration in man on awakening from methoxyflurane, halothane, ether, and fluroxene anesthesia. Anesthesiology 1970;33:5–9
6. Knill RL, Clement JL. Variable effects of anaesthetics on the ventilatory response to hypoxaemia in man. Can Anaesth Soc J 1982;29:93–99
7. Stoelting RK, Eger E I II. Percutaneous loss of nitrous oxide, cyclopropane, ether and halothane in man. Anesthesiology 1969;30:278–283
8. Dale O, Brown BR Jr. Clinical pharmacokinetics of the inhalational anesthetics. Clin Pharmacokinet 1987;12:145–167

9. Targ AE, Yasuda N, Eger E I II. Solubility of I-653, sevoflurane, isoflurane, and halothane in plastics and rubber composing a conventional anesthetic circuit. Anesth Analg 1989;69:218–225
10. Eger E I II, Strum DP. The absorption and degradation of isoflurane and I-653 by dry soda lime at various temperatures. Anesth Analg 1987;66:1312–1315
11. Fink BR. Diffusion anoxia. Anesthesiology 1955;16:511–519
12. Sullivan SF, Patterson RW. Posthyperventilation hypoxia: theoretical considerations in man. Anesthesiology 1968;29:981–986
13. Jones RM. Desflurane and sevoflurane: inhalation anaesthetics for this decade? Br J Anaesth 1990;65:527–536
14. Holaday DA, Smith FR. Clinical characteristics and biotransformation of sevoflurane in healthy human volunteers. Anesthesiology 1981;54:100–106
15. Quasha AL, Eger E I II, Tinker JH. Determination and applications of MAC. Anesthesiology 1980;53:315–334
16. Cahalan MK, Johnson BH, Eger E I II. Relationship of concentration of halothane and enflurane to their metabolism and elimination in man. Anesthesiology 1981;54:3–8
17. Cromwell TH, Eger E I II, Stevens WC, Dolan WM. Forane uptake, excretion and blood solubility in man. Anesthesiology 1971;35:401–408
18. Eger E I II, Johnson BH. Rates of awakening from anesthesia with I-653, halothane, isoflurane, and sevoflurane. Anesth Analg 1987;66:977–982
19. Eger E I II. Isoflurane (Forane): a compendium and reference. Madison, Wisconsin: Ohio Medical Products, 1981
20. Clark NJ, Meuleman T, Liu W-S, et al. Comparison of sufentanil-N_2O and fentanyl-N_2O in patients without cardiac disease undergoing general surgery. Anesthesiology 1987;66:130–135
21. Milligan KR, Howe JP, Dundee JW. Halothane and isoflurane in outpatient anaesthesia. Anaesthesia 1988;43:2–4
22. Chinyanga HM, Vandenberghe H, MacLeod S, et al. Assessment of immediate post-anaesthetic recovery in young children following intravenous morphine infusion, halothane, and isoflurane. Can Anaesth Soc J 1984;31:28–35
23. Sessler DI, Israel D, Pozos RS, et al. Spontaneous post-anesthetic tremor does not resemble thermoregulatory shivering. Anesthesiology 1988;68:843–850

Hemodynamic Consequences of Emergence

Aryeh Shander, MD
Shubhinder Puri, MD

The gratification of completing a complicated anesthetic procedure is enhanced by the safe transfer of the patient from the operating room to the Postanesthesia Care Unit (PACU). The process of returning patients to their previous conscious state requires skill, experience, and an ever-expanding core of knowledge that addresses the patient's underlying medical condition and its interaction with surgery and anesthesia.

In this chapter, the general principles of emergence hemodynamics will first be addressed, followed by a listing of the factors encountered during emergence that influence or interfere with hemodynamic homeostasis. Although not yet a separate entity, the parameters governing the hemodynamic consequences of emergence may one day be approached as a specific topic, and modification of the "emergence response" may be targeted through the use of specific pharmacological agents.

Although reduction of anesthesia-related morbidity and mortality has been enhanced through the use of new monitoring devices and techniques and new drugs [1, 2], one area requiring attention is the upgrading of monitoring protocols for the transport of patients from the operating room to the PACU.

The major determinants of emergence hemodynamics are:

1. The preoperative medical condition of the patient
   a) Cardiac diseases—Coronary artery disease, congestive heart failure, valvular disease, pericardial disease, conduction abnormalities, and arrhythmias [3]
   b) Hypertension—Essential and secondary [4, 5]
   c) Pulmonary diseases—Obstructive, restrictive, parenchymal, infectious, and pulmonary vascular diseases
   d) Any other disease or syndrome that alters the basic homeostatic physiology of the patient causing hemodynamic derangement, including trauma, renal, neurological, endocrine and metabolic conditions

2. The surgery
    a) The nature of the surgical procedure
    b) Location and length of incision
    c) Duration of the surgical procedure
    d) The relationship of specific types of surgery to emergence, such as tracheal extubation of the neurosurgical patient
3. The anesthetic technique and any complications [6]

■ Emergence

The restoration of all vital functions to normal—including ventilation, airway reflexes, and overall hemodynamics—is the goal of emergence. The process of emergence can be broken up into the following four phases:

1. Reduction in depth of anesthesia with recovery of sensory and motor functions
2. Resumption of spontaneous ventilatory control
3. Return of airway reflexes
4. Awakening

Although emergence is begun as the surgical procedure ends, planning for emergence must begin earlier—in many instances prior to surgery. For some critically ill patients, invasive monitoring, as well as continuation of mechanical ventilation into the postoperative period [7], must be contemplated in order to promote a smooth emergence. In all cases, planning for postoperative pain management must be accomplished preoperatively or intraoperatively to attenuate the "emergence response."

Emergence techniques are numerous, but they can be roughly divided into two categories:

1. Rapid emergence, best tolerated by healthy patients undergoing uncomplicated surgery, especially outpatients
2. Deliberately slow and prolonged emergence, preferred for fragile patients with multiple medical problems undergoing prolonged or complicated surgery and anesthesia

Regardless of the technique employed, the anesthesiologist must always try to prevent periods of inadequate anesthesia, amnesia, or analgesia—which tend to occur if surgical and anesthetic events are not well coordinated [8].

Successful emergence also depends on the anesthesiologist's appreciation of the pharmacokinetics and pharmacodynamics of the agents used. Ignoring the effects of small concentrations of these drugs, which can linger in the body for long periods of time, can increase the incidence of complications.

Of all the events that occur during emergence, the riskiest to the patient is extubation of the trachea. Once the patient has regained sufficient motor function and medullary initiation of ventilation, the trachea may be extubated with the patient either in the awake state or still under the influence of anesthetic agents. Either technique may be employed, depending upon the circumstances. For example, to avoid stimulation of the cough reflex or the sympathetic nervous system, the patient may be extubated "deep." An alternative method of suppressing these reflexes while allowing the patient to awaken involves the use of narcotics, local anesthetics, or cardiovascular depressants (e.g., adrenergic blockers, calcium channel blockers, and nitroglycerin) [9]. These techniques, as with attempts to block the response to tracheal intubation, may not deliver the desired results. Ultimately, it is left to the anesthesiologist's clinical skill and judgment to protect the patient's airway while blunting the hemodynamic response to emergence.

■ Hemodynamics

The causes of impaired cardiovascular stability during emergence from anesthesia are many and varied. Although most hemodynamic problems resolve before they become critical and require intervention, preparation for handling them demands a thorough understanding of physiological principles.

Cardiac performance depends on adequately perfused and functioning heart muscle, valves, conduction system, and autonomic nervous system. For the peripheral circulation and the organs to receive oxygenated blood, the heart must deliver sufficient output. Many factors besides muscle function influence cardiac output [10]—defined as heart rate times stroke volume. The stroke volume (end diastolic volume − end systolic volume) is a function of cardiac contractility, preload, and afterload. Thus, the four major determinants of cardiac output are heart rate, preload, afterload, and contractility [10].

The heart rate is controlled by the autonomic nervous system and is altered by a variety of pharmacological agents. *Preload* can be defined in terms of the length of the myocardial muscle fiber prior to contraction. This is usually referred to as the volume present in the ventricles at the end of diastole (left or right end diastolic volume). The ventricular wall stress developed by systolic contraction defines *afterload*. The shape, volume, and thickness of the ventricular wall, as well as the pressure gradient the muscle contracts against, all affect afterload. Quantitative expression of afterload is the systemic vascular resistance (SVR), calculated by the formula

$$\frac{\text{MAP} - \text{CVP}}{\text{CO}} \times 80,$$

where MAP = mean arterial pressure, CVP = central venous pressure, and CO = cardiac output. With respect to contractility, in the normal state of cardiac performance the inotropic state of the heart is governed by the autonomic nervous system via catecholamines. These agents and their pharmacological counterparts stimulate myocardial contraction. Anesthetics, other drugs such as sympatholytics, and conditions that adversely affect oxygen delivery or utilization to the heart have a profound negative effect on myocardial contractility [11].

Arterial blood pressure measurement is the parameter most widely used to determine hemodynamic performance. However, it is becoming increasingly apparent to clinicians that blood pressure measurement, although invaluable, does not reliably reflect individual organ perfusion. Evaluating the determinants of blood pressure—SVR and cardiac output—helps in assessing the state of the vasculature but, even so, cannot define the blood flow or oxygen delivery needed to maintain organ function [12, 13].

Perfusion of organs may be estimated by measurement of mean arterial pressure (MAP). Still, MAP is only one of the determinants of organ perfusion, which also depends upon autonomic control of organ vasculature under normal conditions (autoregulation). If autoregulation is left to operate without interference, O_2 delivery to organ tissues is dependent only on cardiac output and the O_2-carrying capacity of the blood.

Hemodynamic derangement is encountered during emergence from both regional and general anesthesia. Although the following discussion centers on general anesthesia, many of these considerations apply to both. Impaired hemodynamics usually result from the following:

1. Alteration in blood pressure—Hypertension or hypotension, including hypoperfusion secondary to myocardial ischemia and/or cardiac arrhythmias
2. Hypoxemia and hypercarbia
3. Altered temperature states (hypothermia or hyperthermia)
4. Intravascular volume changes (hypervolemia or hypovolemia)
5. Residual drug effects and reactions
6. Influence of the surgical procedure

Blood Pressure and Emergence

The most frequently encountered cause of hemodynamic instability during all phases of emergence is hypertension. Patients with underlying hypertension may develop wide fluctuations in blood pressure during surgery and anesthesia, which can be attributed to their hyperresponsive autonomic nervous system or abrupt discontinuation of antihypertensive drugs in the preoperative period. Other common causes of elevated blood pressure during emergence which can appear in the otherwise normotensive patient include pain, hypoxemia, hypercarbia, agitation, shivering, and residual sympathomimetic stimulation from drugs used during surgery. Blood pressure elevation in this setting results from stimulation of the autonomic nervous system and augmentation of circulating catecholamine levels.

In healthy individuals, marked elevation in blood pressure may be well tolerated for short periods of time and usually resolves spontaneously. Therefore, once underlying causes have been ruled out in this group of patients, treatment should be approached judiciously. In contrast, although patients with chronic hypertension may seem to tolerate sustained elevations in blood pressure, treatment of these patients should be instituted more promptly to minimize end-organ damage. In the compromised cardiac patient, severe elevations in blood pressure increase afterload with a resultant increase in myocardial O_2 requirements. If this need cannot be met, the new imbalance between O_2 supply and demand will lead to further cardiovascular deterioration [14].

Systemic blood pressure elevation normally evokes compensatory mechanisms (i.e., baroreceptor reflexes) to slow heart rate and reduce cardiac output. During emergence, these responses are blunted or nonfunctioning due to residual anesthetic effects, and blood pressure elevations are sustained [15].

Therapies for hypertension are numerous and beyond the scope of this chapter. However, when treatment of hypertension is contemplated during emergence, the cause must be properly identified so that blood pressure elevation secondary to pain is not treated solely with beta blockers or other antihypertensive medications.

Hypotension

The causes of hypotension during emergence can be divided into three major categories:

1. Reduced preload
2. Depressed myocardial performance
3. Decreased SVR

Insufficient fluid repletion or inadequate blood loss replacement in the perioperative period results in reduced preload. When the intravascular deficit reaches a critical level, hypotension ensues. Additionally, some anesthetic drugs, such as narcotics, enflurane, and nitroglycerin, increase venous capacitance and, in so doing, produce hypotension or markedly enhance the hypotensive effects from hypovolemia.

A fall in cardiac output due to depressed myocardial performance can lead to hypotension. This reduced inotropic function can be a result of multiple factors such as preexisting congestive heart failure, myocardial ischemia, and dysrhythmias. Hypotension may be aggravated in the presence of reduced preload by any negative inotropic influence.

A reduction in SVR is an uncommon cause of hypotension during emergence from general anesthesia. Reduced SVR occurs secondary to vasodilation due to sepsis, hyperthermia, or sympathectomy of a surgical or chemical nature (e.g., spinal anesthesia), or to the development of large peripheral arteriovenous shunts.

Therapeutic intervention to correct hypotension should be directed toward the specific underlying cause. Because the cause may not be readily apparent, maneuvers to maintain organ perfusion must be undertaken while attempts to identify the source of hypotension are sought.

Myocardial ischemia can cause hemodynamic embarrassment, either by generating a low perfusion state due to depressed muscle function or by precipitating malignant dysrhythmias that in turn reduce effective ventricular filling and ejection. The high catecholamine state observed during early phases of ischemia may be a consequence of either the pain it produces or the low flow state resulting from pump failure. Increased levels of circulating catecholamines may exacerbate myocardial ischemia and perpetuate the cardiac dysfunction and/or dysrhythmias [10].

The etiology of dysrhythmias during anesthesia and emergence involves several factors and is influenced by the myocardial cellular environment [16].

Dysrhythmias that result in hemodynamic compromise should be treated promptly. For example, patients with compromised cardiac function and atrial fibrillation, whose decreased ventricular filling is due either to losing their atrial "kick" or to developing a rapid ventricular rate with short diastolic filling times, develop a low output state and require intervention. Ventricular tachyarrhythmias that produce myocardial ischemia, severe hypotension, and reduced cardiac output also demand immediate attention.

Hypoxia and Hypercarbia

Both hypoxia and hypercarbia are common events during emergence. (The respiratory causes are discussed elsewhere in this issue.) The severity varies with the situation and with duration but can have devastating effects on the cardiovascular system. Both hypoxia and hypercarbia can produce blood pressure elevation, tachycardia, or bradycardia, and can also directly depress the myocardium resulting in dysrhythmias and/or impaired organ perfusion.

Hypercarbia is common following cessation of mechanical ventilation when attempts to stimulate spontaneous respiration are begun. Both a narcotic technique and inhalation anesthesia elevate the respiratory center apneic threshold. The rate of rise of CO_2 is dependent on many factors during emergence; if rapid, it may cause decreased alveolar oxygen tension (P_AO_2) in the lungs and hence with the circulation. Sympathetic stimulation occurs secondary to hypercarbia, and elevation in heart rate and blood pressure ensue with deleterious effects on the cardiovascular system.

Usually, by the time hypoxia effects cardiovascular instability, this complication has gone unnoticed for many minutes. When ventilatory support is withdrawn, or when the trachea is extubated, frequent inspection for early signs of hypoxia or hypercarbia is important.

Alteration in Temperature

Much attention has been directed to understanding the physiological implications of temperature control and its relationship both to overall homeostasis and to cardiopulmonary function. Although deliberate hypothermia offers some advantage in preserving myocardium and brain tissue [17], development of hypothermia during routine operations can have serious deleterious consequences during emergence. Hypothermia depresses the myocardium directly, and also indirectly by precipitating characteristic arrhythmias such as bradycardia, premature ventricular contractions, ventricular tachycardia, and ventricular fibrillations. Hypothermia also increases the work of the heart by raising SVR secondary to vasoconstriction and increased blood viscosity. Hypothermia, when severe, renders most pharmacological agents, particularly sympathomimetics, ineffective by interfering in reuptake or metabolism of these drugs. The activity of these agents may become more apparent during the rewarming phase of emergence.

The shivering encountered during emergence may be due to either hypothermia or inhalation anesthetics [18]. The end result is an acute rise in O_2 demand which may lead to myocardial decompensation and its deleterious effects on the circulation [19]. Passive rewarming using available modalities such as fluid warmers, airway humidity traps or warmers, warming blankets, and increased environmental temperature are usually sufficient to prevent major heat loss. Ascertaining the patient's temperature should be a PACU priority and can be done while treatment with passive rewarming is begun.

Although not common, hyperthermia in the PACU should always raise the clinician's suspicion of malignant hyperthermia. More common causes, however, include intercurrent infection, sepsis, blood product or drug reaction, and inadvertent overwarming of the patient during surgery. Rises in temperature result in increased heart rate, increased cardiac output and O_2 consumption, and tax a poorly functioning heart, leading it to fail.

Disturbances in Volume Status: Anemia, Hypovolemia, and Hypervolemia

The delivery of oxygen to tissues depends on the hemoglobin concentration, the oxygen saturation of hemoglobin, and the cardiac output. Changes in any of these variables alter the capability of the system to operate optimally. The influence of hemoglobin concentration is paramount, because small changes result in large swings in oxygen-carrying capacity, which ultimately affects oxygen delivery. There is recent evidence that, under defined circumstances (e.g., absence of severe cardiopulmonary disease), a hemoglobin of 7 g/dl or more is sufficient. Anemia of this magnitude does not interfere with wound healing or the patient's recovery and may in fact improve perfusion to ischemic areas by altering rheology [20].

Impaired hemodynamics secondary to anemia usually derive from compensatory cardiovascular mechanisms. The reduced supply of oxygen to the periphery during anemia is met by increasing cardiac output. Because the stroke volume tends to be reduced by hypovolemia, tachycardia becomes the operant variable. In addition to this physiological response, the outpouring of catecholamines increases SVR, which tends to maintain blood pressure and preferential organ perfusion [21].

Complicated by reduced oxygen delivery, an imbalance of oxygen supply and demand may become a problem during emergence for patients with underlying cardiopulmonary disease. In such cases, any factor that increases the oxygen requirement during emergence, e.g., shivering, agitation, or pain, aggravates myocardial ischemia and reduces cardiac output. Particular attention should be paid to patients with impaired hemodynamics and to those with a decreased capability for oxygen unloading (as with hypothermia, alkalosis, 2,3-diphosphoglyceric acid deficiency, or methemoglobinemia). For patients with marginal hemodynamics these imbalances may be accentuated by the stress of emergence.

Complications of hypervolemia include hypertension, congestive heart failure, pulmonary edema, cerebral edema, and, with free water overload, associated electrolyte disturbances. Hypervolemia in the perioperative period follows excess intravenous fluid replacement or free water infusion during hysteroscopy or transurethral resection of the prostate.

Drug Effects

Most anesthetics that produce myocardial depression have dissipated to minimal residual concentrations by the late stages of emergence, but may continue to have important pharmacological effects. Some pharmacological agents that have known negative cardiovascular effects, such as protamine [22] and antibiotics, may be administered specifically at the end of surgery. These drugs, when infused rapidly, release histamine-inducing vasodilation and pulmonary bronchoconstriction. These effects are poorly tolerated by an already decompensated heart [23].

Other drugs that affect hemodynamic stability are the narcotic antagonists and anticholinesterases such as physostigmine used to "reverse" delirium. When naloxone is used to antagonize the respiratory depression of narcotics during anesthetic emergence, large doses may cause catecholamine release and stress the cardiovascular system. Titration of narcotic antagonists to analgesic requirement may help restore adequate spontaneous respiration without undue tachycardia or hypertension. Anticholinesterase agents function as neuromuscular blockade reversal agents; tertiary amine varieties that cross the blood-brain barrier can reverse the somnolence of general anesthesia, theoretically by increasing centrineural cholinergic transmission. Any anticholinesterase drug can precipitate bradyarrhythmias. In general, drug effects and reactions (such as anaphylaxis) are numerous and must constantly be kept in mind as possible causes of unexplained hemodynamic deterioration or collapse [23].

■ Surgical Site and Procedure

Certain surgical procedures place added demands on anesthetic management during emergence. Neurosurgical procedures in particular require the patient to emerge with minimal cardiovascular perturbation, yet themselves exert profound effects on the cardiovascular system.

Protracted intubation of the trachea as anesthesia lightens is difficult to achieve without inducing a sympathetic response, resulting in tachycardia and elevation of blood pressure, and occasionally precipitating cardiac arrhythmias and pulmonary edema. Therapeutic intervention to obtund the "emergence response" has not been uniformly successful. Adrenergic blockers, calcium channel blockers, nitrates, and local anesthetics have all been tested and have not been found to be effective in all situations. When appropriate, however, any of the above regimens, alone or in combination, may help. Planning ahead for pain control, i.e., patient-controlled analgesia, peridural narcotics, or other analgesics, aids in attenuating this response to emergence.

Positioning of neurosurgical patients postoperatively also affects their hemodynamics. Patients who have undergone craniotomies may be placed in a 30° upright position, which, compounded by hypovolemia secondary to diuretic therapy and/or bleeding, may lead to hypotension and tachycardia. Another example of cardiovascular instability can be seen in the patient emerging hypertensive, following carotid surgery; less commonly, hypotension may occur.

Airway irritation from an endotracheal tube, due to head manipulation during head and neck surgery, as well as aspiration of oral secretions during extubation, can lead to sustained coughing. Forceful coughing can raise arterial blood pressure, increase venous congestion, and cause bleeding in patients who have undergone surgery in highly vascularized areas, jeopardizing the surgical outcome.

Transporting the Emerging Patient from the Operating Room to the PACU

Intrahospital transportation of critically ill patients carries great risk. In the absence of a specific study addressing this risk, we have assumed, perhaps correctly, that it has been reduced in the past two decades through the use of physiological monitoring during transport [24, 25].

Despite recommendations for continuous monitoring during transport from the operating room, in general we tend to do what seems familiar to us. Stable respiratory and hemodynamic status and the ability of an extubated patient to protect the airway are usually sufficient criteria for transfer to the PACU [8]. At this point clinical evaluation replaces physiological monitoring.

Concomitant with the building of larger hospitals and larger operating suites, and with increased use of anesthetics outside the operating room, distances from anesthetizing sites to PACU and the time required for transport have increased.

The hazards of intrahospital transport have been recognized for some time [25] and include hypertension, hypotension, arrhythmias, hypoxia [26, 27], hypercarbia acidosis, airway obstruction, and cardiac arrest. Though this area needs more study, some clinicians have begun to better control this phase of emergence by implementing "transport monitoring." Questions regarding its cost/benefit relationship will have to be answered by controlled studies, but the feeling of the authors is that such monitoring should become a standard of care [28]. The parameters to be monitored should include, as a minimum, blood pressure, electrocardiogram, pulse oximetry, and breath sounds. In addition to monitoring, supplemental oxygen administration is likely also to become a transport standard of care.

The above recommendations are not new to individuals who deal with transport of critically ill patients. The availability of special "transport" beds equipped with attached and readily accessible monitors and oxygen supply will help make implementation universal.

Hemodynamics and Transport

The period of emergence is usually brief, and complications are limited by the short time that patients are exposed to this phase. However, moving patients can stimulate tachycardia and hypertension [29]. Superimposed pain only compounds the problem. The previously stable patient who becomes agitated or hypoxic (appearing cyanotic) during the course of transfer requires rapid transport to the PACU, which in itself may contribute to hemodynamic instability by shifting intravascular volume during rapid turns around sharp corners [30]. Patients who are vulnerable to temperature change may lose additional heat during transport, further contributing to hemodynamic instability.

■ Summary

Attempts to attenuate the hemodynamic "emergence response" depend upon the availability and use of "real-time" physiological monitors. Such devices increase our diagnostic acuity and accuracy during this critical and labile period. Future investigation is needed to determine the effects of monitoring on morbidity and mortality, and is likely to lead to the development of guidelines and standards of care for the transport of emerging patients from the operating room to the PACU.

■ References

1. Eichhorn JH. ASA panel examines improved anesthesia morbidity and mortality. APSF 1990;5:1–12
2. Tinker JH, Dull DL, Caplan RA, et al. Role of monitoring devices in prevention of anesthetic mishaps. Anesthesiology 1981;71:541–546
3. Goldman L. Cardiac risk and complications of noncardiac surgery. Ann Intern Med 1983;98:504
4. Asiddao CB, Donegan JH, Whitesell RC, et al. Factors associated with perioperative complications during carotid endarterectomy. Anesth Analg 1982;61:631
5. Merli GJ, Weitz HH. Preoperative consultation. Med Clin North Am 1987;71: 399–412; 413–432
6. Goldman L, Caldera DL, Nussbaum SR, et al. Multifactorial index of cardiac risk in non-cardiac surgical procedures. N Engl J Med 1977;297:845–850
7. Pinsky MR. The influence of positive-pressure ventilation on cardiovascular function in the critically ill. Crit Care Clin 1985;1:699–717
8. Willenkin RL. Management of general anesthesia. In: Miller RD, ed. Anesthesia. 3rd ed. New York: Churchill/Livingstone, 1990:1343–1344
9. Kamra S, Wig J, Sapru RP, et al. Topical nitroglycerine: a guard against pressor responses to tracheal intubation. Anaesthesia 1986;41:1087–1091
10. Williams GA, Ayres SM. Regulation of myocardial function in health and critical illness. Crit Care Clin 1985;1:435–451
11. Filner BE, Karliner JS. Alterations of normal left ventricular performance by general anesthesia. Anesthesiology 1976;45:610–621
12. Kreuzer F, Cain SM. Regulation of the peripheral vasculature and tissue oxygenation in health and disease. Crit Care Clin 1985;1:453–470
13. Bland RD, Shoemaker WC. Common physiologic patterns in general surgical patients: hemodynamic and oxygen transport changes during and after operation in patients with and without associated medical problems. Surg Clin North Am 1985;65:793–809
14. Attia RR, Murphy JD, Snider M, et al. Myocardial ischemia due to infrarenal aortic cross-clamping during aortic surgery in patients with severe coronary artery disease. Circulation 1976;53:961
15. Kotorly KJ, Ebert TJ, Vucins E, et al. Baroreceptor reflex control of heart rate during isoflurane anesthesia in humans. Anesthesiology 1984;60:173
16. Atlee JL III, Zeljko JB. Mechanism for dysrhythmias during anesthesia. Anaesthesia 1990;72:347–374

17. Swan H. The importance of acid-base management for cardiac and cerebral preservation during open heart operations. Surg Gynecol Obstet 1984;158:391–414
18. Pozos RS, Isreai D, McCutchenn R, et al. Human studies concerning thermal-induced shivering, postoperative "shivering," and cold-induced vasodilation. Ann Emerg Med 1987;16:1037–1041
19. Guffin A, Girard D, Kaplan JA, et al. Shivering following cardiac surgery: hemodynamic changes and reversal. J Cardiothorac Anesth 1987;1:24–28
20. Tuma RF, White JV, Messmir K. The role of hemodilution in optimal patient care. Munich, Germany: Zuckschwerdt, 1989:21–41; 42–53
21. Waxman K, Shoemaker WC. Physiologic response to massive intraoperative hemorrhage. Arch Surg 1982;117:470–475
22. Shapira N, Schaff HV, Piehler JM, et al. Cardiovascular effects of protamine sulfate in man. J Thorac Cardiovasc Surg 1982;84:504
23. Stoelting RK. Allergic reactions during anesthesia. Anesth Analg 1983;62:341
24. Wright I, Rogers PN, Ridley S. Risks in intrahospital transport (letter). Ann Intern Med 1987;8:638
25. Braman S, Dunn S, Amico C, Millman P. Complications of intrahospital transport in critically ill patients. Ann Intern Med 1987;107:469–473
26. Pullerits J, Burrows FA, Roy WL, et al. Postoperative complications: hypoxia during transfer to the recovery room; arterial desaturation in healthy children during transfer to the recovery room. Can J Anaesth 1987;34:470–473
27. Gervais HW, Eberle B, Konietzke D, et al. Comparison of blood gases of ventilated patients during transport. Crit Care Med 1987;15:761–763
28. Eichhorn JH, Cooper JB, Cullen DJ, et al. Standard for patient monitoring during anesthetic at Harvard Medical School. JAMA 1986;256:1017–1020
29. Insel J, Weissman C, Kemper M, et al. Cardiovascular changes during transport of critically ill and postoperative patients. Crit Care Med 1986;14:539–542
30. Waddell G. Movement of critically ill patients within hospital. Br Med J 1975;2:417–419

Respiratory Problems in Emergence from Anesthesia

Richard L. Austin, Jr, MD

Anesthesiologists are frequently involved with respiratory problems in the recovery room, ranging from a simple mucus plug obstructing an endotracheal tube to fulminant pulmonary edema from overwhelming sepsis. As the severity of illness increases in many institutions, and as more and more operative procedures are performed on an outpatient basis, anesthesiologists are faced with managing sicker patients for longer periods in the recovery room. In this discussion of postoperative pulmonary problems, I will first review some of the alterations in pulmonary function known to result from general anesthesia and mechanical ventilatory support, and then proceed to some of the common respiratory difficulties encountered in postoperative care.

Anesthesia and Respiratory Function

It is well established that general anesthesia results in major changes in pulmonary gas exchange [1]. It is documented that, during spontaneous ventilation, the respiratory depressant effects of either inhaled or intravenous agents may result in alveolar hypoventilation. As tidal volume decreases, dead space ventilation (rather than alveolar ventilation) comprises a greater percentage of the tidal volume—i.e., the dead space–to–tidal volume ratio (V_D/V_T) increases. Following upper abdominal surgery, whether due to pain, decreased pulmonary compliance, or altered diaphragmatic function, patients breathe in a faster, shallower pattern [2]. Should tidal volume decrease to such an extent that it approaches anatomical dead space, the potential for arterial hypoxemia is obvious.

It is quite common, however, even in the mechanically ventilated patient receiving larger tidal volumes, for decreases in arterial oxygenation during general anesthesia and emergence to occur. The expected arterial oxygenation during general anesthesia at a specified oxygen concentration is obtained by applying the alveolar air equation $P_AO_2 = FI_{O_2}(Pb - P_{H_2O}) - P_ACO_2/RQ$, where FI_{O_2} = fractional inspired concentration of oxygen; Pb = barometric pressure (normal = 760 mm Hg); P_{H_2O} = water vapor pressure (normal = 47 mm Hg); P_ACO_2 = alveolar partial pressure of CO_2; and RQ = respiratory quotient (normal = 0.8).

Thus, for a patient receiving 50% oxygen by face mask, and assuming a normal P_ACO_2 and respiratory quotient, we would expect: $P_AO_2 = 0.50 (760 - 47) - 40/0.8$. Or, $P_AO_2 = 356 - 50$, or 306.

In the absence of important right-to-left intrapulmonary shunt, arterial Po_2 should approach alveolar Po_2. A large discrepancy between alveolar and arterial oxygen tensions (alveolar-arterial gradient) indicates pulmonary pathology. A Po_2 of 120 mm Hg, then, for a patient breathing 100% oxygen is quite adequate for oxygen transport purposes, but at the same time may be a harbinger of such difficulties as imminent pulmonary edema or massive alveolar collapse.

One possible cause for a widened alveolar-arterial (A-a) gradient during anesthesia and early emergence is inhibition of hypoxic pulmonary vasoconstriction by all potent inhalational agents and many commonly used vasodilators [3]. In the unanesthetized pulmonary vasculature, pulmonary arterioles are known to constrict in an area of poorly ventilated lung, thus redistributing pulmonary artery outflow to better ventilated regions. In addition to inhalational agents, nitroglycerin, nitroprusside [4], dobutamine [4], and isoproterenol have been reported to blunt hypoxic pulmonary vasoconstriction. This inhibition of hypoxic pulmonary vasoconstriction may be responsible for some of the early decreases in P_AO_2 postoperatively, but later and more persistent decreases in oxygenation demand other explanations.

A-a gradients, although useful, have several limitations. As mentioned above, abolition of hypoxic pulmonary vasoconstriction is a common cause for a widened A-a gradient. In addition to the anesthetic agents and vasodilators, high concentrations of oxygen can also increase the magnitude of right-to-left intrapulmonary shunt. Ironically it seems apparent, then, that the common practice of determining A-a gradients on 100% oxygen leads to overestimation of the true degree of pulmonary pathology. In addition, A-a gradients do not recognize stagnant or anemic etiologies for arterial hypoxemia. Should oxygen demand exceed oxygen supply, desaturation of mixed venous blood may occur. In the presence of right-to-left intrapulmonary shunting, the addition of desaturated venous blood to left ventricular output may contribute to arterial hypoxemia. A more accurate assessment, therefore, of the true degree of pulmonary pathology is reflected in the use of the shunt equation

$$\frac{\dot{Q}_S}{\dot{Q}_T} = \frac{C_c'O_2 - C_aO_2}{C_c'O_2 - C_vO_2}$$

where $\dot{Q}_S \backslash \dot{Q}_T$ = intrapulmonary shunt fraction; $\dot{Q}_S$ = flow through shunt; $\dot{Q}_T$ = cardiac output (L/min); $C_c'O_2$ = idealized pulmonary capillary O_2 content, = (1.34 ml/g Hg × Hg (g/dl) × SaO_2) + (0.003 × P_AO_2); C_aO_2 = systemic arterial O_2 content, = (1.34 ml/g Hg × Hg (g/dl) × SaO_2) + (0.003 × PaO_2); C_vO_2 = mixed venous O_2 content, = (1.34 ml/g Hg × Hg (g/dl) × SvO_2) + (0.003 × PvO_2).

The primary advantage of the shunt equation over use of A-a gradients is the incorporation of mixed venous oxygen levels in determining the magnitude of intrapulmonary pathology. It should be remembered that there are stagnant, anemic causes for hypoxemia that may contribute to arterial hypoxemia. Lack of appreciation of this fact may result in errors of therapy by, for example, treating arterial hypoxemia with positive end-expiratory pressure (PEEP) rather than recognition that an inotrope is needed to correct severe venous desaturation due to cardiogenic shock.

■ Mechanical Ventilation and Respiratory Function

In the mechanically ventilated patient, physiological dead space and intrapulmonary shunt increase as a result of altered intrapulmonary gas distribution. With the many recent advances in computerized diagnostic imaging, our understanding of gas distribution within the chest continues to progress from the pioneering diaphragmatic fluoroscopy described by Froese and Bryan [5]. Several subsequent studies have confirmed their observation of a cephalad displacement of the diaphragm in anesthetized patients during controlled or spontaneous ventilation. Hedenstierna and colleagues [6] have estimated the magnitude of this loss of lung volume at 0.5 L after a 1.9-cm cephalad shift. Thus, the absence of active contractile diaphragmatic tone during controlled ventilation seems to be the primary event responsible for loss of lung volume leading to basilar atelectasis and potential arterial hypoxemia. Although the diaphragm has been the central focus of investigation, computed tomography has also shown a reduction of cross sectional chest area, causing a further 0.3-L loss in lung volume during controlled ventilation. Hedenstierna and co-workers speculate that the alterations in diaphragmatic contractile tone may initiate the chest wall–rib cage changes. Tensing the diaphragm is purported to expand the rib cage, whereas diaphragmatic relaxation may cause reduction of the rib cage diameter.

Just as important as the absolute, quantitative changes in lung volume are qualitative, distributional changes in intrapulmonary gas distribution. During spontaneous ventilation in anesthetized patients, displacement of the diaphragm is substantially greater in dependent lung areas, as opposed to that in nondependent areas. During mechanical ventilation, however, a greater proportion of a delivered mechanical breath is delivered to anterior, nondependent areas. During passive diaphragmatic displacement, motion is described as "piston-like," and gas is preferentially distributed to nondependent (anterior, in supine patients) areas when relatively low tidal volumes are delivered. Larger tidal volumes partially compensate for the lack of diaphragmatic displacement and to some extent deliver gas to dependent lung areas.

The mechanisms responsible for these changes remain controversial. Froese and Bryan [5] offer a vertical abdominal hydrostatic gradient as the primary factor responsible for the anterior (nondependent) distribution of ventilatory breaths, i.e., the diaphragm becomes progressively more difficult to displace caudally in more dependent (posterior) areas of lung, primarily due to the greater weight of abdominal contents impinging on the diaphragm posteriorly.

Another possible explanation for the anterior distribution of mechanically delivered breaths is the tendency of air delivered under pressure to follow the path of least resistance. Pleural pressures in the supine patient are more negative anteriorly, gradually approaching atmospheric pressures posteriorly. This gradient accounts for the presence of larger alveolar and small airway diameters anteriorly. Additionally, posterior areas of the lung (zone 3) are highly perfused, and pulmonary vasculature may encroach on airspace diameter. The combination of more negative pleural pressures anteriorly and little encroachment by the pulmonary vasculature on the corresponding airspaces may ease the delivery of air under pressure to anterior airways of larger diameter.

Literally turning Froese's argument upside down, Krayer and associates [7] failed to find a vertical abdominal hydrostatic gradient in mechanically ventilated patients in the prone position. Abdominal contents hang freely in the prone patient, and it may be that in the prone patient, as in the upright patient, there is little force counteracting diaphragmatic displacement. It seems likely that there are other factors at work in addition to an abdominal hydrostatic gradient, but they remain to be elucidated.

Consistent with the preceding discussion, the clinical repercussions of mechanical ventilation relate to an increase in physiological dead space ventilation and a tendency for loss of lung volume posteriorly. Accordingly, anesthetized or recovering patients should receive larger tidal volumes than the 6 to 8 ml/kg of the awake, unanesthetized patients. Most recommendations for intraoperative and postoperative ventilatory management include tidal volumes of 10 to 15 ml/kg. Routine use of capnography has led to the realization that many patients receiving large tidal volumes and the conventional 10 breaths/min have substantial respiratory alkalosis. Rather than decreasing tidal volumes to respond to this new information, it may be prudent, instead, to decrease the number of breaths to 6 to 8/min.

The loss of functional residual capacity (FRC) is thought to be the primary insult responsible for the increased physiological right-to-left intrapulmonary shunt resulting in decreases in oxygenation during anesthesia and recovery. Consequently, attempts to maintain FRC offer advantages in oxygenation during emergence. Following surgery not involving the upper abdomen, the extent of dead space ventilation and the tendency for posterior lung volume loss (leading to shunt, hypoxemia, atelectasis, and possible pneumonia) should theoretically be lessened by having the patient, to as great a degree as possible (short of inspiratory muscle fatigue), breathe spontaneously, or augment mechanical ventilation with intermittent spontaneous breathing. Mechanical ventilation following upper abdominal or thoracic procedures involves other considerations, which will be discussed shortly. Although assist-control mechanical ventilatory modes give the illusion of patient augmentation of ventilation, only the initial triggering of ventilation is spontaneous, and motion of the diaphragm is still passive and pistonlike. Intermittent mandatory ventilation (IMV) is the only currently available, widely used mode offering major patient contribution to ventilation. In addition to better gas exchange and distribution, IMV offers improved cardiovascular performance, increased renal perfusion, lessened incidence of barotrauma, avoidance of respiratory alkalosis, and better patient acceptance.

■ Respiratory Function after Abdominal and Thoracic Procedures

Thus far, we have discussed primarily the effects of anesthesia and mechanical ventilation per se on lung function. Separating the effects of anesthesia and mechanical ventilation from those of the surgical procedure, particularly in upper abdominal and intrathoracic sites, has been a recurrent difficulty in interpretation of clinical studies. Upper abdominal procedures are associated with a 55% decrease in vital capacity and a 30% reduction in FRC [7]. This deficit is thought to persist for a period of 10 days to 2 weeks before spirometric measures of lung function return to preoperative levels. The observation that measured respiratory function rapidly returns to normal following anesthesia and peripheral surgery suggests that it is the site of the surgical incision itself, and not changes induced by the administered anesthetic, that is at issue. Further support emerges from the fact that lower abdominal procedures cause changes in pulmonary function that are intermediate in severity between upper abdominal or thoracic procedures and those involving the periphery.

Decreases in forced vital capacity (FVC) and forced expiratory volume in 1 second (FEV_1) are consistently observed immediately following upper abdominal or thoracic procedures, whereas some investigators have not found substantial reductions in FRC until 16 hours postoperatively [8]. It is important to consider these alterations in lung volume separately, recognizing that each has different clinical implications, rather than grouping them together as "abnormal pulmonary function tests." FEV_1 is probably the best indicator of a patient's ability to clear secretions postoperatively, and the ratio of FEV_1 to FVC is a sensitive indicator of expiratory outflow obstruction. A normal postoperative FVC correlates with full neuromuscular recovery and the ability to defend against hypoventilation and respiratory acidosis.

FRC, though seldom measured clinically, is the single most important determinant of the patient's ability to maintain oxygenation. FRC is the resting, end-expiratory lung volume toward which most postoperative lung expansion maneuvers are directed. Progressive decreases in FRC may cause tidal breathing to cross below the lung volume at which dependent alveoli tend to occlude, i.e., closing volume. In this situation, alveoli become unventilated, gas absorption continues along gradients toward the microvasculature, and alveolar collapse ensues. Meyers and colleagues [9] have noted that reduction of FRC in excess of 40% is notable radiographically as atelectasis. However, the absence of atelectasis radiographically does not preclude the possibility that substantial "microatelectasis" is present. The relative insensitivity of conventional radiographs for years hampered efforts to determine how great a role airway closure and atelectasis played in hypoxemia during anesthesia and emergence. Again aided by more sophisticated imaging techniques, Tokics and co-workers [10] demonstrated by computed tomography the rapid development of pulmonary densities, interpreted as atelectatic areas, shortly following anesthesia induction with spontaneous ventilation. Interestingly, the size of these densities was noted to increase during mechanical ventilation, even though mechanical ventilation during anesthesia is frequently preferred by practicing anesthesiologists (as opposed to spontaneous ventilation), due to concerns about alveolar hypoventilation and collapse. These studies seem to confirm the conclusions of Froese and Bryan relating to hypoventilation of posterior, dependent lung fields during mechanical ventilation. There seems little doubt that airway closure does indeed occur during both mechanical and spontaneous ventilation in anesthetized patients, and that it contributes to inefficient oxygenation [11]. During postoperative recovery in the mechanically ventilated patient, dependent atelectasis probably continues for the duration of mechanical ventilation. How long atelectasis persists in the extubated, spontaneously breathing patient following surgery not involving the abdomen or thorax is unanswered, but the large number of outpatients safely discharged from operative suites without benefit of incentive spirometry, respiratory therapists, or postoperative education suggests a rapid return to normal lung volumes.

Alveolar collapse may be hastened by high levels of inspired oxygen; the lack of the usually occurring nitrogen alveolar "stent" that provides internal support after oxygen is absorbed into the pulmonary vasculature causes "absorption atelectasis." This condition is believed to occur as alveoli filled with oxygen become emptied of all gas and subsequently collapse.

■ Pain and Respiratory Function

Although it seems obvious that postoperative pain plays a major role in decreased lung volume after upper abdominal or thoracic procedures, until recently there were few good studies demonstrating the benefits of postoperative analgesia for preserving respiratory function. Within the past few years, however, Bromage and colleagues [12], Welchew and Thornton [13], Modig [14], Rawal and co-workers [15], and Yeager and associates [16] have all demonstrated major advantages to the use of epidural narcotics or local anesthetics in relation to postoperative pulmonary function. Bromage and colleagues [12] demonstrated considerable improvements in both FEV_1 and FVC, as well as in levels of postoperative sedation. In a study of 53 patients scheduled for high-risk surgery, Yeager and co-workers [16] showed definite improvements in duration of intubation, pulmonary fluid overload, hospital cost, and perioperative mortality. The differences were so striking that the authors felt it necessary to terminate their study and offer epidural pain relief to all their high-risk patients.

Postoperative epidural narcotics may have some advantage over the use of epidurally administered local anesthetics, inasmuch as local anesthetics can impair motor function. One disadvantage of epidural narcotics is the widely acknowledged importance of observing such patients in an intensive care setting, due to concerns about cephalad migration of epidural narcotics and subsequent respiratory depression. With the increasing use of pulse oximetry and capnography, particularly with remote telemetry (analogous to EKG monitoring from a central location), greater numbers of patients may be afforded the advantages of epidural analgesia for longer periods without compromise of patient safety.

Other advantages of postoperative epidural analgesia, besides better respiratory function, include a lower incidence of thromboembolic events [17], decreased intraoperative blood loss, a decrease in perioperative catabolism, better flow through surgically placed vascular grafts, and, as previously mentioned, earlier ambulation and less need for postoperative sedatives.

Although effective pain management is central to improving respiratory function after surgery, other factors are involved. In the postoperative patient who is "unwilling or unable" to inspire deeply, it is unclear whether the primary problem is unwillingness to breathe deeply secondary to pain, or truly an inability to inspire deeply due to anatomical or physiological pathology.

The residual effects of general anesthetics or muscle relaxants following surgery or administration of postoperative narcotics can produce alveolar hypoventilation; narcotics also eliminate sighing ventilatory patterns. In addition to these alterations in central respiratory drive, respiration is affected by dysfunction of the diaphragm itself in the postoperative period. By directly stimulating the phrenic nerve following upper abdominal surgery, Dureuil and associates [18] found that the intrinsic ability of the diaphragm to contract was well preserved postoperatively. However, Ford and coauthors [19] and Dureuil and colleagues [18] postulate a reflexive inhibition of phrenic nerve output via afferent vagal, splanchnic, or sympathetic receptors, resulting in decreased central phrenic nerve output and, consequently, hampered ventilation.

Aminophylline has been shown to have a beneficial effect on diaphragmatic function after upper abdominal surgery [20]. Its known effect on central respiratory centers and its effectiveness in the management of both neonatal and adult obstructive sleep apnea further support the existence of a central nervous component to diaphragmatic dysfunction. A direct agonist effect of aminophylline on diaphragmatic contractility is also possible.

■ Common Postoperative Respiratory Problems

Upper Airway Obstruction

The most common cause of postoperative respiratory compromise is upper airway obstruction. Identifying characteristics of upper airway obstruction are either noisy or silent inspiration and the presence of paradoxical chest wall motion. The obstruction may be due to relaxation of the musculature of the upper airway with posterior displacement of the tongue, resulting in occlusion of the posterior pharynx. This is best managed by anterior manipulation of the mandible, allowing the jaw to jut forward, with simultaneous hyperextension of the neck to further open the upper airway. Should obstruction persist, an oral or nasal airway (or both if necessary) should be placed.

In the shivering, jaw-clenching, awakening patient, placement of an oral airway may be very difficult. Additionally, should the patient with masseteric trismus vomit without an oral airway in place, suctioning of the oropharynx may be difficult, making pulmonary aspiration of gastric contents more likely. It is advisable, therefore, to have an oral airway in place before emergence begins. Certainly there are hazards involved with the use of an oral airway, including precipitation of gagging and vomiting, dental trauma, and laryngospasm, but these complications can usually be avoided by timely removal of the airway as the patient awakens. Bite blocks are effective for protecting the teeth and may have fewer adverse effects, but they do nothing to provide an unobstructed airway.

High-pitched inspiratory sounds in the presence of a clear oropharynx usually indicate the presence of laryngospasm or upper airway obstruction secondary to airway edema. Tracheal extubation during light levels of anesthesia may provoke laryngospasm, but, if performed when the patient is awake and responsive, seldom results in dangerous vocal cord closure. Laryngeal obstruction following thyroidectomy may be secondary to bilateral recurrent laryngeal nerve transection or, even less commonly, severe hypocalcemia. More frequently, upper airway edema results from multiple attempts at tracheal intubation. Upper airway obstruction may also occur following laryngeal surgery, extrinsic compression from bleeding after thyroid surgery, or extensive head and neck procedures for cancer. Rarely, upper airway edema results from systemic anaphylaxis; other signs of anaphylaxis, however, such as hypotension, wheezing, or urticaria are usually present. Laryngospasm or obstruction due to edema can usually be acutely managed by continuous positive airway pressure (CPAP) using a face mask. As obstruction from edema or laryngospasm continues, increasingly negative airway pressure develops, causing soft tissues of the upper airway to swell and further encroach on an already narrowed or obstructed airway. CPAP probably exerts its major benefit by preventing this secondary detrimental response, allowing upper airway pressures to normalize and preventing the inward collapse of compliant upper airway structures.

Should CPAP by mask not rapidly improve upper airway obstruction, other measures must be promptly initiated. If the presumptive cause of obstruction is laryngospasm, intravenous sedation and administration of succinylcholine is indicated. When edema is the working diagnosis, and when time permits, inhaled racemic epinephrine, elevation of the head, opening of the surgical incision, or glucocorticoids may all be useful. In the presence of upper airway edema, the approach to reintubation should be carefully considered, as the administration of succinylcholine may distort the airway and render the trachea difficult to intubate.

A particular form of acute, fulminant, pulmonary edema has now been well described following acute upper airway obstruction. In Willms and Shure's review [21], laryngospasm was the most common cause. Markedly negative intrathoracic and transpulmonary pressures, brought on by the patient's attempt to breathe in against a closed glottis, was thought to result in egress of fluid from pulmonary capillaries to the interstitial space. Other possible mechanisms include central volume overload due to increased venous return in the presence of negative intrathoracic pressure, a hyperadrenergic state also causing increased central blood volume, or hypoxic pulmonary vasoconstriction producing high pulmonary capillary pressures.

Pulmonary Edema

Pulmonary edema is one of the more common problems facing the anesthesiologist in the postoperative period and may be generally categorized as either cardiogenic or noncardiogenic in nature. Cardiogenic pulmonary edema is usually the result of either excessive amounts of fluid administered intraoperatively or of left ventricular dysfunction causing a sequestration of fluid in the lungs. Edema fluid is usually low in protein content, due to the preserved integrity of the vascular endothelium of the lungs. Pulmonary artery occluded pressures are high and cardiac output is usually depressed. It is rare in young, healthy individuals for intravenous fluid administration to be sufficiently great as to overwhelm the ability of pulmonary lymphatics to clear edema fluid, as well as the ability of cardiovascular and renal mechanisms to excrete the excess fluid. If pulmonary edema results, there is usually concomitant cardiovascular or renal insufficiency, although it may have been well compensated and unappreciated prior to surgery.

Several mechanisms may be responsible for the development of pulmonary edema in the perioperative period. Salt and water retention due to increased levels of antidiuretic hormone (ADH) and aldosterone is a normal response of the hypothalamus and pituitary to the stress of surgery. Mechanical ventilation also promotes increased secretion of ADH. Myocardial depressant effects of potent inhalational agents may also play a role in decreased renal perfusion. The result in either case is oliguria in the face of rising pulmonary artery pressures.

Without any readily available monitor, other than urinary output, to ensure adequate renal perfusion, the practicing anesthesiologist understandably responds to oliguria by continuing fluid administration. However, if oliguria persists, invasive hemodynamic monitoring to ensure adequate cardiac performance is warranted. Despite the risk of overhydration, it is still true that a little pulmonary edema is easier to manage than a little renal failure.

Another factor responsible for the development of postoperative pulmonary edema is the management of intraoperative hypotension due to the vasodilating properties of general anesthetics, and, particularly, spinal or epidural anesthesia, by infusion of large quantities of crystalloid solutions. As vasoconstriction returns during the postoperative period, fluid is redistributed from the periphery to the central vasculature. Early in the postoperative course, patients may experience further intense peripheral vasoconstriction secondary to catecholamines in response to either hypothermia or pain. A key question is how much fluid to administer to patients with preexisting cardiac or renal insufficiency undergoing spinal anesthesia, in the face of small intraoperative losses? An even more problematic scenario is the simultaneous presence of large intraoperative losses and extensive sympathectomy.

Finally, the literature now extensively reflects the hazards of indiscriminate use of the opioid antagonist naloxone. Sudden reversal of the venodilating effects of opioids, as well as an acute increase in systemic vascular resistance secondary to pain, may produce fulminant pulmonary edema. It is my practice to avoid the use of naloxone in the context of anesthesia for patients with cardiac disease. Certainly, there are indications for the use of narcotics in patients with impaired left ventricular function, but a patient ill enough to remain somnolent or apneic after anesthesia including narcotics need not be extubated soon after termination of the operative procedure, and naloxone should not be given to awaken the patient or promote breathing.

Noncardiogenic pulmonary edema, while less common, may present a more difficult challenge for the anesthesiologist. Pulmonary artery occluded pressures are low, and edema fluid is rich in protein, reflecting the permeability defect present in the vascular endothelium. Common causes for noncardiogenic pulmonary edema include sepsis, aspiration of gastric contents, a fat embolus, pulmonary contusion, and near-drowning. Whether massive transfusion or shock produces a permeability defect in the lung remains unclear.

The great majority of cases of noncardiogenic pulmonary edema seen by the anesthesiologist are attributable to sepsis. The characteristic hemodynamic profile of these patients includes low mean arterial pressure, low systemic vascular resistance, and high cardiac output. In contrast to most other forms of respiratory failure, initial arterial blood gases usually reflect not a respiratory acidosis, but alkalosis. Early in the course of the disease, the pulmonary defect is not one of ventilation but of oxygenation. The oxygenation deficit has been characterized as "refractory hypoxemia" in that it responds poorly to increases in fractional inspired oxygen concentration.

Once again, decreased FRC is the primary physiological insult responsible for hypoxemia. Pulmonary compliance is severely reduced, and very high inflationary pressures may make delivery of selected tidal volumes impossible using conventional anesthesia ventilators. Because prolonged administration of oxygen, by itself, produces a permeability defect, and because high concentrations of oxygen are relatively ineffective, most therapeutic maneuvers are directed toward restoration of FRC toward normal. This goal is accomplished through the use of either CPAP administered by mask or, more commonly, by positive end expiratory pressure (PEEP) and mechanical ventilation using an endotracheal tube. PEEP not only restores an acceptable level of oxygenation but may also improve pulmonary compliance and reduce the work of breathing. The use of PEEP may appreciably reduce the number of mechanically delivered breaths needed; the use of CPAP may obviate the need for tracheal intubation entirely.

Both CPAP and PEEP have a substantial deleterious impact on cardiovascular performance, however, and, just as important, on the interpretation of invasively obtained hemodynamic data. PEEP may reduce cardiac output in a number of ways. Decreased venous return secondary to increases in intrathoracic pressure has long been appreciated. Also, inasmuch as pulmonary artery hypertension, an early hallmark of evolving noncardiogenic pulmonary edema, may produce right ventricular dysfunction, PEEP may lower cardiac output by increasing impedance to right ventricular ejection. It is not uncommon for large areas of the lung to be relatively unaffected by the pulmonary insult, and for PEEP administered to the entire lung to overinflate unaffected subpopulations of alveoli. These overinflated alveoli may then compress the adjacent pulmonary vasculature, worsening right ventricular performance.

Equally important is the effect of PEEP on interpretation of hemodynamic data. Should the pulmonary artery catheter lie in an area of the lung with a high $\dot{V}/\dot{Q}$ ratio (zone 1), the catheter may function as an airway pressure monitor as much as a monitor of intravascular pressures. Care should be taken to estimate filling pressures at end-expiration. With increasing pulmonary artery hypertension, right ventricular overload and dilation begin to encroach on left ventricular chamber size by leftward displacement of the interventricular septum. (Here may be the ultimate example of pressure and volume not correlating in linear fashion!) While the left ventricular end-diastolic pressures are rising due to encroachment within the pericardium from the right, left ventricular end-diastolic volume is being reduced.

In patients with severe noncardiogenic pulmonary edema from septic shock (where the mortality rate approaches 50%) ultimate outcome may hinge on the quick resolution of pulmonary edema. Inasmuch as the primary cause of hypotension is lowered systemic vascular resistance, and that many patients with hypotension lack evidence of ischemic heart disease, it seems reasonable to consider early use of vasoconstrictors. Thus, because very small increases in pulmonary hydrostatic pressures may result in large increases in interstitial lung water, "optimization" of preload to a filling pressure of 18 to 20 mm Hg may be inappropriate. Close attention should be directed toward indices of oxygen transport, including cardiac output, mixed venous oxygen content, and arteriovenous oxygen content difference.

Should cardiac output be compromised due to higher levels of positive pressure ventilation, early intervention with inotropic support, rather than continued volume challenge, may be appropriate in this patient population. Furthermore, in patients with an ongoing endothelial permeability leak, there seems to be little advantage in administering colloid solutions, which ultimately find their way to the pulmonary interstitium. Colloid administration may have a role later in the course of the disease, as the permeability defect is repaired and continued low vascular oncotic pressures make resolution of the remaining interstitial water difficult.

■ Summary

In conclusion, anesthesia and mechanical ventilation have major effects on respiratory function, both intraoperatively and postoperatively. Recognition that mechanical ventilation represents a major departure from spontaneous ventilation should enable anesthesiologists to compensate for the increases in dead space ventilation and the propensity for alveolar collapse that accompany low volume mechanical ventilation. The use of postoperative regional analgesic techniques to alleviate respiratory compromise constitutes both current clinical practice and an area of active, ongoing investigation. Gone are the days when the anesthesiologist's responsibility stopped at the recovery room door. Today, anesthesiologists with expertise in postoperative pain management, cardiovascular physiology, and intensive ventilatory care are able to provide patients an improved likelihood of avoiding postoperative respiratory complications.

■ References

1. Rehder K, Sessler AD, Marsh HM. State of the art: general anesthesia and the lung. Am Rev Respir Dis 1975;112:541–563
2. Ali J, Weisel RD, Layug AB, et al. Consequences of postoperative alterations in respiratory mechanics. Am J Surg 1974;128:376–382
3. Mathers JM, Benumof JL, Wahrenbrock EA. General anesthetics and regional hypoxic pulmonary vasoconstriction. Anesthesiology 1977;46:111–114
4. Marshall BE, Marshall C. Anesthesia and the pulmonary circulation. In: Covino BG, Fozzard HA, eds. Effects of anesthesia. Bethesda: American Physiological Society, 1985:121
5. Froese AB, Bryan CH. Effects of anesthesia and paralysis on diaphragmatic mechanics in man. Anesthesiology 1974;41:242–255
6. Hedenstierna G, Strandberg A, Brismar B, et al. Functional residual capacity, thoracoabdominal dimensions, and central blood volume during general anesthesia with muscle paralysis and mechanical ventilation. Anesthesiology 1985;62:247–254
7. Krayer S, Rehder K, Vettermann J, et al. Position and motion of the human diaphragm during anesthesia-paralysis. Anesthesiology 1989;70:891–898
8. Ali J, Weisel RD, Layug AB, et al. Consequences of postoperative alterations in respiratory mechanics. Am J Surg 1974;128:376–382

9. Meyers JR, Lembeck L, O'Kane H, Baue AE. Changes in functional residual capacity of the lung after operation. Arch Surg 1975;110:576–583
10. Tokics L, Hedenstierna G, Strandberg A, et al. Lung collapse and gas exchange during general anesthesia: effects of spontaneous breathing, muscle paralysis, and positive end-expiratory pressure. Anesthesiology 1987;66:157–167
11. Rehder K, Marsh HM, Rodarte JR, Hyatt RE. Airway closure. Anesthesiology 1977;47:40–52
12. Bromage PR, Camporesi E, Chestnut D. Epidural narcotics for postoperative analgesia. Anesth Analg 1980;59:473–480
13. Welchew EA, Thornton JA. Continuous thoracic epidural fentanyl. Anaesthesia 1982;37:309–316
14. Modig J. Respiration and circulation after total hip replacement surgery. Acta Anaesthesiol Scand 1976;20:225–236
15. Rawal N, Sjostrand U, Christoffersson E, et al. Comparison of intramuscular and epidural morphine for postoperative analgesia in the grossly obese: influence on postoperative ambulation and pulmonary function. Anesth Analg 1984;63:583–592
16. Yeager MP, Glass DD, Neff RK, Brinck-Johnsen T. Epidural anesthesia and analgesia in high-risk surgical patients. Anesthesiology 1987;66:729–736
17. Modig J, Borg T, Karlstrom G, et al. Thromboembolism after total hip replacement: role of epidural and general anesthesia. Anesth Analg 1983;62:174–180
18. Dureuil B, Viires N, Cantineau JP, et al. Diaphragmatic contractility after upper abdominal surgery. J Appl Physiol 1986;61:1775–1780
19. Ford GT, Whitelaw WA, Rosenal TW, et al. Diaphragm function after upper abdominal surgery in humans. Am Rev Respir Dis 1983;127:431–436
20. Dureuil B, Desmonts JM, Mankikian B, Prokocimer P. Effects of aminophylline on diaphragmatic dysfunction after upper abdominal surgery. Anesthesiology 1985;62:242–246
21. Willms D, Shure D. Pulmonary edema due to upper airway obstruction in adults. Chest 1988;94:1090–1092

Complications during Recovery

Roger S. Mecca, MD

Persistent Sedation

Successful progression through the differential diagnosis of prolonged unconsciousness after general anesthesia requires an organized approach. First, it is useful to assess the level of preoperative responsiveness to rule out preexisting cerebral dysfunction or the possibility of unrecognized drug or alcohol intoxication. Time and amount of all preoperative and intraoperative medications with sedative properties or potentiating side effects should be noted, and any intraoperative events should be thoroughly reviewed.

Examination of an obtunded patient in the Postanesthesia Care Unit (PACU) includes careful checking of vital signs. In addition to establishing the presence of spontaneous ventilation, evaluating its rate and character is of special importance since degree of ventilatory depression relates to residual anesthetic depth. Evaluation of heart rate and rhythm and systemic blood pressure can indicate adequacy of cerebral perfusion and the prevailing level of autonomic tone. Pupillary size and response should be evaluated, although the diagnostic value of these findings is questionable. Assessment of level of consciousness should include a firm tactile stimulus if a verbal stimulus fails to elicit arousal.

Residual sedation from anesthetic medications is the most frequent cause of postoperative somnolence [1]. Use of long-acting sedatives (e.g., pentobarbital, hydroxyzine, promethazine, droperidol, lorazepam, scopolamine) for premedication or supplementation of anesthetic agents will contribute to postoperative somnolence. Prolonged unconsciousness from inhalation anesthetics is more likely after long surgical procedures (especially in obese patients) or when high inspired concentrations of inhalation anesthetics are continued through the end of surgery to facilitate a smooth "deep" extubation. Sedation caused by intraoperative narcotic or sedative administration is generally dose-related. After a typical anesthesia course, arousal occurs within 60 to 90 minutes of admission to the PACU, even in patients who are susceptible to sedation.

If unconsciousness persists, low-dose intravenous naloxone (0.04-mg increments every 2 minutes up to 0.2 mg) can be administered in an attempt to reverse the sedative effect of intraoperative narcotics. Careful titration of naloxone allows reversal of respiratory depression and sedation without precipitating dangerous reversal of analgesia and extreme hypertension and tachycardia [2]. If a significant increase in spontaneous ventilation and arousal are not noted after 0.2 mg of intravenous naloxone, it is unlikely that unconsciousness is related to residual narcotic effect unless massive overdose has occurred. Administration of intravenous physostigmine (1.25 mg) sometimes counteracts sedation caused by inhalation anesthetics and other sedative medications, probably through nonspecific, acetylcholine-mediated central nervous system (CNS) arousal [3, 4]. Administration of flumazenil, a new benzodiazepine antagonist, may prove useful to reverse the central sedative effects of midazolam and diazepam [5, 6]. In patients anesthetized under emergency conditions, it is possible that an unrecognized preoperative overdose with depressant oral medication is responsible for prolonged sedation, even when no response is seen after administration of naloxone and physostigmine.

On rare occasions, profound residual neuromuscular paralysis might mimic unconsciousness in the PACU by precluding a noticeable response to stimuli. This could occur after gross overdosage with nondepolarizing relaxants, if reversal agents are inadvertently omitted or in patients with unrecognized neuromuscular disease, as well as with phase II block caused by pseudocholinesterase deficiency or excessive succinylcholine administration. Observation of spontaneous ventilation, purposeful motion, reflex activity, or other evidence of neuromuscular function eliminates residual paralysis as an explanation for unresponsiveness.

Once residual effects of anesthetic agents or neuromuscular relaxants have been ruled out, various other causes of postoperative unresponsiveness should be evaluated. Patients who were exhausted preoperatively will often be difficult to rouse after a general anesthetic. This is especially true in children having emergency surgery at night when normal sleep routines have been interrupted. Body temperature should be assessed, since hypothermia below 33°C can directly impair consciousness and accentuate the potency and duration of depressant medications. (At core temperatures below 30°C, fixed dilation of pupils, absence of reflexes, and progressive evolution of coma can occur.) Evaluation of serum glucose will eliminate severe hypoglycemia or hyperglycemic, hyperosmolar coma as causes. Suspicion that unresponsiveness may be caused by hypoglycemia is an indication for an immediate empirical trial of intravenous 50% dextrose solution. Serum electrolyte concentrations and osmolarity should also be checked to rule out iatrogenic hypoosmolar states such as acute hyponatremia. Arterial blood gas analysis is useful to assess pulmonary function, because severe hypercarbia and carbon dioxide narcosis from prolonged hypoventilation can cause coma, as can unrecognized hypoxemia.

If the above analysis does not yield a diagnosis, a thorough neurological evaluation should be performed with careful documentation of findings. Consultation with a neurologist is usually helpful at this point. An occasional patient suffering subclinical grand mal seizures, secondary to an underlying seizure disorder, or delirium tremens will present with unresponsiveness in the PACU. CNS depression secondary to intravenous local anesthetic toxicity or inadvertent subarachnoid injection can mimic postoperative coma. Untoward intraoperative events such as severe hypotension, dysrhythmias, hypoxemia, or hypercarbia must be considered as potential causes of cerebral anoxia during surgery. In injured patients, one must consider the possibility of unrecognized head trauma and increasing intracranial pressure. Intraoperative cerebral thromboembolism is another possibility, especially in patients recovering from cardiac, proximal major vascular, or invasive neck surgery [7, 8], or in those who have undergone internal jugular, subclavian, or intraarterial cannulation. Patients with a history of atrial fibrillation, symptomatic carotid occlusion, or hypercoagulable states are also at increased risk of thromboembolism [9]. Postoperative cerebrovascular accidents in other patients are rare and usually occur later in the postoperative course [10]. Paradoxical air embolism through a right-to-left intracardiac shunt, or intracerebral hemorrhage secondary to intraoperative hypertension can also cause postoperative cerebrovascular accidents [11]. Of course, in patients recovering from intracranial surgery, increased intracranial pressure from bleeding, edema, or pneumocephalus must be considered [12]. Finally, it is always possible that a patient is feigning unresponsiveness.

■ Altered Mental Status

On occasion, a patient recovering from a major anesthetic will manifest an acceptable level of arousal but abnormal mental reactions ranging from confusion and lethargy through extreme disorientation and physical combativeness. Establishing an etiology for such reactions and dealing with them present a difficult challenge for the entire PACU staff.

Delirium and confusion during emergence from anesthesia should be viewed as having the same import as any other untoward postoperative event. Aside from the disturbance and upset the situation might cause to staff and other patients in the PACU, there are real risks associated with this problem. Agitated patients often manifest high levels of sympathetic nervous system tone with consequent tachycardia and hypertension, each of which can cause serious medical complications. Forceful, thrashing movements of combative patients can jeopardize suture lines, orthopedic fixations, vascular grafts, drains, tracheal tubes, and indwelling vascular catheters, thus increasing the risk of postoperative morbidity. The risk of incidental trauma also increases, raising the possibility of contusion or fracture from contact with equipment or side rails, corneal abrasion from dislodged oxygen apparatus, and sprains from violent struggling against restraints. Least appreciated is the risk of injury to PACU staff grappling to contain a combative patient.

Reactions to emergence from general anesthesia are most often psychological. For some time after restoration of consciousness there is a lack of integration of higher cerebral functions that interferes with the patient's ability to process and react to sensory input. This lack of integration takes various forms. The majority of emerging patients are quiet, sleepy, and slightly disoriented, with sluggish mental reactions that clear gradually. Some experience wide emotional swings, most often reflected by uncontrolled weeping during emergence. On occasion, patients manifest marked combativeness with escalating physical resistance to positioning and restraint.

It is difficult to predict which patients will have unusual reactions on emergence. Certainly those with preoperative personality aberrations would be expected to exhibit those aberrations during emergence. There is a higher frequency of emergence problems in patients with mental retardation, clinically evident psychiatric disorders, organic brain dysfunction, or hostile interactions preoperatively. Ethnic, cultural, and individual personality differences among patients undoubtedly play some role [13], probably related to perceived acceptable modes of dealing with pain and expressing discomfort. A language barrier can accentuate an emergence reaction, because reassurance from PACU staff may not be understood. Inability to communicate caused by maxillary fixation or endotracheal intubation can also generate frustration and fear that will exaggerate emergence reactions. Extreme emergence reactions are more prevalent in children and younger adults than in the elderly. Anxiety in young children is undoubtedly heightened by separation from the parents. The incidence of stormy emergences is probably higher after surgical procedures that are emotionally charged (e.g., breast or testicular biopsies).

Patients taking medications for psychogenic disorders, or those premedicated with long-acting sedatives, may exhibit clouded sensoria and disorientation postoperatively. Certainly patients who abuse alcohol, narcotics, barbiturates, cocaine, or other drugs might exhibit bizarre behavior during emergence caused by acute preoperative intoxication or unrecognized withdrawal symptoms. In addition, adverse psychological reactions with disorientation and combativeness have been described after use of parenteral scopolamine as a premedication or antiemetic. Scopolamine-induced disorientation can be treated with intravenous physostigmine [14]. Delirium induced by other anticholinergics is usually less severe, but patients receiving long-term preoperative meperidine therapy or atropine premedication may also exhibit postoperative delirium [15, 16]. Ketamine has also been implicated in postoperative dysphoria and hallucination; however, adverse reactions to ketamine are rare, especially in the PACU setting. Use of etomidate as an induction agent can contribute to postoperative restlessness [17]. Untoward psychological reactions to a host of other medications have been anecdotally reported.

Regardless of an individual patient's predisposition to emergence reactions, the presence of noxious stimuli during emergence certainly amplifies confusion, agitation, and aggressive behavior. Because the most common noxious stimulus is surgical pain, every effort should be made to ensure adequate postoperative analgesia early in the PACU course. Other unpleasant stimuli can also exacerbate an emergence reaction. Urinary bladder or gastric distention often causes pain and agitation in emerging patients. Tight dressings, painful phlebotomy, or poor positioning worsen agitation, as does discomfort from endotracheal or nasogastric tubes, urinary catheters, or infiltrated vascular catheters. One should always check for unrecognized sources of discomfort such as small pieces of equipment inadvertently left beneath a patient, corneal abrasions, or entrapment of sensitive body parts.

Patients may also react to noxious sensations that are not necessarily painful. Nausea and dizziness are very distressing to an emerging patient, as is severe pruritus caused by narcotics or other medications. Some patients are far more comfortable in a semi-sitting position and will struggle vigorously during emergence to sit up, particularly obese patients or those with symptomatic gastroesophageal reflux or pulmonary congestion. Inability to move can generate considerable agitation during emergence. Patients often fight vigorously against physical restraint imposed by personnel or binders, only to become much quieter once restraint is relaxed. Inadequate recovery from neuromuscular relaxation commonly causes severe anxiety and agitation during emergence, even in the absence of ventilatory compromise. Unexplained weakness is terrifying and sometimes prompts violent, uncoordinated motions that can make a patient appear disoriented and combative. The patient's lack of strength and the characteristic flapping movement are helpful in making the diagnosis of residual neuromuscular blockade—as is quantitative electrical nerve stimulation.

Confusion, delirium, or combativeness after anesthesia may also reflect serious dysfunction of major organ systems. Inadequate respiratory function is often the problem. Moderate hypoxemia in emerging patients can produce clouded mentation, disorientation, and agitation that are difficult to distinguish from pain reactions. Respiratory acidemia also causes profound agitation and discomfort, especially if hypercarbia is related to airway obstruction or poor ventilatory mechanics. (Hypercarbia secondary to ventilatory center depression generates somewhat less agitation because higher CNS functions are also depressed.) Hypercarbia without coincident respiratory acidemia is generally asymptomatic unless the arterial partial pressure of carbon dioxide reaches such high levels that early carbon dioxide narcosis causes somnolence and disorientation.

Postoperative agitation may also reflect perceived inadequate ventilation. Limitation of tidal volume by tight chest dressings, increased intra-abdominal pressure, gastric distention, splinting against a painful incision, or decreased pulmonary compliance gradually causes a vague dissatisfaction with lung inflation similar to that of air hunger. This problem is common in patients receiving postoperative mechanical ventilation with low delivered volumes and is probably mediated by stretch receptors which respond to changes in lung and chest wall volume. Inability to generate a forceful cough or to clear secretions can also be distressing, as can labored breathing from increased airway resistance or partial upper airway obstruction. Agitation related to these phenomena can be profound even though ventilation and oxygenation appear adequate by arterial blood gas analysis.

Certain cardiovascular problems accentuate anxiety during emergence from anesthesia. Engorgement of pulmonary vasculature with early interstitial pulmonary edema can cause symptoms of chest fullness and air hunger before airway flooding occurs. Subjective symptoms of frequent ventricular ectopy are also distressing, especially if staff members allow the patient to see their concern about the EKG findings. If cardiac output or systemic pressure is inadequate to maintain peripheral perfusion, lactic acidemia can result, causing anxiety and mild disorientation. When systemic blood pressure falls so low that cerebral perfusion is not maintained, patients may exhibit lethargy, severe disorientation, agitation, and combativeness. This situation constitutes a medical emergency that requires aggressive restoration of cardiac output or peripheral vascular resistance. Inappropriate administration of sedative or analgesic medications in such patients could generate a catastrophic cerebral event.

Several metabolic abnormalities can interfere with lucidity in the postoperative period. Acute hyponatremia after transurethral prostatic resection is a widely discussed example of a hypoosmolar state that markedly clouds sensoria. Similar acute cerebral fluid shifts should be considered in dialysis patients, cases of massive fluid infusion, and after acute intravascular repletion following severe dehydration. Acute hyperosmolar states probably cause more subtle mental changes postoperatively. Hyperglycemia from excessive glucose infusion or insufficient insulin could cloud consciousness during recovery if hyperosmolarity were severe. Although uncommon, severe hypoglycemia can cause agitation or markedly diminished responsiveness during recovery.

Once these common causes of delirium or agitation have been ruled out, one must consider the possibility of a primary neurological problem. Unrecognized seizure activity in the postoperative period can cause profound changes in mental status. The seizure itself can mimic agitation and combativeness, or the patient can appear disoriented, somnolent, and confused during the postictal phase. One should be particularly wary of seizures in patients with a history of head trauma, chronic alcohol intoxication, cocaine abuse, or epilepsy. Acute cerebral embolism, hemorrhage, or infarct sometimes presents with changes in orientation, ability to vocalize, or level of consciousness [18].

Therapy for altered mental status in the PACU is essentially supportive and varies with etiology. For emergence reactions, waiting for residual anesthesia to wear off is usually sufficient. Most uncomplicated emergence reactions improve within 10 to 15 minutes. Verbal reassurance that surgery has been completed and that the patient's status is acceptable can be invaluable. Staff can help reestablish orientation by using the patient's and surgeon's name frequently and by stressing time and location. It is important to thoroughly assess the patient's complaints no matter how bizarre they may seem, for more often than not there is an element of reality to the patient's feedback. It is often useful to allow a patient to determine his or her own position when feasible. If possible, offending stimuli should be removed or minimized. Certainly one should ensure adequate pain relief. In some patients, small doses of parenteral sedative will relieve fear or anxiety and smooth the emergence. Identifying whether a patient is reacting to pain or to anxiety is important when choosing which medication to employ. Parenteral narcotics are relatively poor sedatives, and benzodiazepines or barbiturates are ineffective analgesics. In the author's opinion, physical restraint by staff or through the use of binding should be used only as a last resort when a patient's physical safety is in jeopardy.

If aggressive evaluation reveals that altered mental status is a symptom of a physiological abnormality (e.g., hypoxemia, acidemia, hypoglycemia, hypotension), sedative or analgesic medications should not be administered. Rather, the underlying problem causing the physiological abnormality should be treated and the patient's mental status repeatedly reassessed as homeostasis is reestablished.

■ Nausea and Vomiting

Postoperative nausea and vomiting are among the most common complications encountered during emergence from general anesthesia, although the incidence varies dramatically with age, duration of anesthesia, surgical procedure, and anesthetic technique [19–23]. Aside from unpleasantness for patient and staff, nausea and vomiting pose genuine medical risks. Increased intra-abdominal pressure or muscle straining can jeopardize abdominal or inguinal suture lines. Transient elevation of central venous pressure resulting from the increased intrathoracic pressure caused by vomiting may increase morbidity after ocular, tympanic, or intracranial procedures. Sympathetic nervous system responses to emesis dramatically elevate heart rate and systemic blood pressure, increasing the risk of cardiovascular complications such as myocardial ischemia or emergence dysrhythmias. Movement during vomiting worsens postoperative pain, accentuating the autonomic response. Gagging and retching can also elicit a parasympathetic response with consequent bradycardia and hypotension. Patients are at obvious risk of aspirating gastric contents during postoperative vomiting, especially if airway reflexes are marginal. After surgical procedures requiring postoperative oral fixation, vomiting and aspiration in the PACU can be catastrophic. Finally, persistent postoperative nausea often delays discharge or necessitates overnight admission of ambulatory surgical patients, decreasing the efficiency of ambulatory services and reducing patient satisfaction with anesthetic care.

Several factors probably contribute to postoperative nausea and vomiting, including starvation, the direct action of anesthetics on chemotactic centers, and autonomic imbalance. Postoperative pain certainly contributes to nausea [24]. The risk of postoperative nausea is higher following surgical procedures involving extraocular muscle or middle ear manipulation, peritoneal or intestinal irritation, and testicular traction [25–28]. Patients with an individual predisposition to nausea (as evidenced by previous postoperative emesis or a history of motion sickness) are also at increased risk. Obesity, hiatal hernia, or other conditions that interfere with the integrity of the gastroesophageal junction probably also increase the likelihood of postoperative emesis. Swallowed blood or secretions promote postoperative vomiting, as does a large accumulation of intragastric gas caused by difficult face mask ventilation, esophageal intubation, or accumulated nitrous oxide. Undergoing general anesthesia near menses may increase the likelihood of vomiting in women [29].

Attempts have been made to implicate individual anesthetic medications or techniques in postoperative nausea and vomiting. The incidence of vomiting is clearly lower after regional anesthesia than after general anesthesia, although the difference becomes blurred when high doses of parenteral narcotics are used to control postoperative pain once a regional block has worn off. Several investigators have noted a decrease in postoperative vomiting when nitrous oxide is avoided [30–32], although the causal connection remains questionable [33–35]. The incidence of postoperative nausea appears relatively constant whether halothane, enflurane, or isoflurane is utilized for inhalation anesthesia [36]. The choice of induction agent affects the incidence of nausea. Barbiturate induction seems less offensive than ketamine or etomidate inductions, and induction with propofol may prove better yet [37]. Preoperative and intraoperative use of narcotic analgesics probably contributes to postoperative nausea when compared with "pure" inhalation techniques [38–41], especially in the ambulatory surgery population. This problem can be partially circumvented by using relatively small doses of the shorter-acting narcotics (e.g., fentanyl 1–1.5 μg/kg, alfentanil 6–8 μg/kg) as part of the induction regimen. Careful titration of postoperative analgesics and supplementation with nonnarcotic analgesics can also decrease the incidence of nausea in ambulatory patients. With respect to longer-acting parenteral narcotics, meperidine seems to generate a higher incidence of postoperative nausea than does morphine.

Although there is no universally recognized prophylactic measure for postoperative nausea and vomiting, several forms of intervention have been evaluated for both prevention and therapy. Providing adequate postoperative analgesia is important, as is care to avoid gastric distention. Prevention of postoperative vestibular stimulation by limiting brisk head motion is also important in this regard. Use of intravenous droperidol as part of the anesthetic regimen has been extensively evaluated [25, 26, 42–46]. Droperidol clearly decreases the incidence and severity of postoperative nausea, although its efficacy varies from study to study and from patient to patient. Intraoperative droperidol can be used effectively in ambulatory patients without unduly prolonging recovery time or causing excessive nausea if the total dosage is kept low (e.g., 1–2 μg/kg, or 1.25 mg total in adults). Repeating intravenous droperidol in the PACU for breakthrough nausea is also effective, although excessive sedation may prolong recovery and delay discharge in ambulatory patients. Care must always be taken to ensure that the alpha-adrenergic blocking properties of droperidol do not precipitate hypotension in hypovolemic patients. Droperidol has been implicated in extrapyramidal symptoms also [47], although these are infrequent, transient, and usually inconsequential.

The use of intravenous metoclopramide for postoperative nausea has also received attention. Administration of metoclopramide either alone [48, 49] or in combination with droperidol [50] has been shown to decrease the incidence of postoperative vomiting in ambulatory patients without prolonging recovery time. Whether the effect of metoclopramide is caused by a central action or by its effects on gastric volume and emptying is unclear. The use of intravenous scopolamine has been largely abandoned because of unacceptable psychogenic reactions during recovery, but a resurgence of interest in scopolamine as an antiemetic occurred when it became available in transdermal form [51–55]. Although transdermal scopolamine may have some benefit as a prophylactic agent, its low efficacy and its tendency to cause postoperative visual disturbances make it a poor choice. Intravenous ephedrine has been described as effective in treating postoperative nausea related to ambulation or motion. Caution should be exercised when using ephedrine in doses advocated by some investigators, because elevation of blood pressure and heart rate may cause more problems than they solve.

Most instances of nausea and vomiting during emergence are self-limited and relate to the surgery and the anesthetics. However, when assessing a patient for nausea and vomiting in the postoperative period, the anesthesiologist should search for other causes such as hypotension, hypoxemia, hypoglycemia, increased intracranial pressure, or gastric bleeding prior to instituting routine types of treatment.

■ Postoperative Acid-base Disorders

Respiratory Acidemia

Potent inhalation anesthetic agents, narcotics, and many sedative medications promote hypoventilation by decreasing CNS sensitivity to lowered cerebrospinal fluid (CSF) pH. It is not surprising, therefore, that respiratory acidemia is perhaps the most frequently encountered postoperative acid-base abnormality. Hypercarbia and acidemia are usually unimportant when the arterial partial pressure of carbon dioxide ranges from 45 to 50 mm Hg in awake, spontaneously breathing PACU patients receiving adequate postoperative analgesia. However, deeply anesthetized or sedated patients can suffer serious acidemia unless supplemental ventilation is instituted or ventilatory depression is reversed.

Postoperative patients are sometimes unable to sustain adequate ventilation in spite of a normal CNS drive to breathe. Increased airway resistance, decreased pulmonary compliance, or residual neuromuscular paralysis can severely impede ventilation and generate progressive respiratory acidemia. An increase in minute carbon dioxide production caused by fever, shivering, hyperalimentation, or malignant hyperthermia also contributes to postoperative respiratory acidemia [56–58]. These problems are especially dangerous in patients with increased baseline respiratory dead space.

Marked postoperative respiratory acidemia causes agitation, confusion, dyspnea, and tachypnea. Increased sympathetic nervous system (SNS) activity manifested as tachycardia, hypertension, and dysrhythmias [59] can increase the risk of myocardial ischemia, postoperative bleeding, or cerebrovascular accident. (When respiratory acidemia is related to CNS depression, signs of SNS activity are often less pronounced because central autonomic responses are depressed as well.) Respiratory acidemia increases cerebral blood flow and intracranial pressure in patients with head injury, intracranial tumors, or cerebral edema. At very low pH levels, catecholamines no longer effectively interact with adrenergic receptors, such that heart rate and blood pressure decrease precipitously.

Acute compensatory mechanisms for respiratory acidemia are limited because the kidneys require many hours to generate compensatory metabolic alkalosis. Therapy for respiratory acidemia must be directed at correcting the imbalance between effective alveolar ventilation and carbon dioxide production. Arousal and stimulation, reversal of neuromuscular relaxants or narcotics, relief of airway obstruction, reduction of airway resistance, and improvement of ventilatory mechanics all improve CO_2 removal. If carbon dioxide elimination cannot be maintained with spontaneous ventilation, then tracheal intubation and mechanical ventilation are necessary. Occasionally, reducing carbon dioxide production by controlling fever or shivering, decreasing the work of breathing, or eliminating high glucose loads from hyperalimentation can be employed. Extreme measures such as core cooling or paralysis are seldom useful in the PACU setting.

Metabolic Acidemia

The possible causes of acute postoperative metabolic acidemia are few. Patients suffering from renal failure or renal tubular acidosis are usually identified by history and almost always exhibit some degree of preoperative metabolic acidemia. An overdose of certain drugs (phenformin, aspirin) and ingestion of toxic substances (methanol) also increase metabolic acid load, but such cases are exceedingly rare. Postoperative metabolic acidemia is occasionally caused by ketoacidosis in severe diabetics. Serum glucose levels are usually elevated and ketones are detectable in blood or urine.

Once preexisting metabolic acidemia and ketoacidosis are excluded, one should assume that postoperative metabolic acidemia is caused by lactic acid accumulation secondary to insufficient delivery or utilization of oxygen in peripheral tissues [60]. In surgical patients, poor peripheral perfusion is often caused by hypotension from low cardiac output (hypovolemia, cardiac failure, dysrhythmia) or decreased systemic vascular resistance (sepsis, catecholamine depletion, sympathectomy). Intense arteriolar constriction from severe hypothermia or inappropriate pressor administration can also reduce perfusion to peripheral tissues, as can poor distribution of blood flow. Hypoxemia, decreased oxygen-carrying capacity of blood (severe anemia, carbon monoxide poisoning), interference with release of oxygen from hemoglobin (alkalemia, hypothermia), or inability of mitochondria to utilize oxygen (cyanide or arsenic poisoning) can also generate lactic acidemia.

The clinical presentation of metabolic acidemia is similar to that of respiratory acidemia; signs and symptoms reflect physiological responses to abnormal pH. However, CNS response to acute metabolic acidemia is often somewhat milder, because hydrogen and bicarbonate ions cross the blood-brain barrier with more difficulty than does carbon dioxide. Similarly, chronic acidemias often generate less dramatic clinical symptoms because CNS bicarbonate concentration adjusts with time to restore brain pH to normal.

In a spontaneously breathing patient with an intact ventilatory drive, generation of respiratory alkalosis to compensate for postoperative metabolic acidemia is rapid [59], but ventilatory responses are markedly attenuated by anesthetic agents [61]. Therapy should be directed at underlying conditions causing accumulation of metabolic acid. Improvement of low cardiac output or hypotension can markedly reduce lactic acid production, as can rewarming. If the acidemia is mild and conditions causing the accumulation of acid are improving, one can allow acidemia to resolve spontaneously through acid metabolism and renal excretion of excess hydrogen ions. However, if acidemia is severe or progressive, intravenous bicarbonate administration is useful to keep the patient's pH near normal.

Respiratory Alkalemia

Hyperventilation caused by excessive operative pain or anxiety during emergence from anesthesia is a common cause of acute respiratory alkalemia in the PACU. Pathological causes of "central" hyperventilation include systemic sepsis, cerebrovascular accident, or paradoxical CNS acidosis (a temporary condition caused by differences in bicarbonate concentration across the blood-brain barrier after prolonged mechanical hyperventilation). Delivery of excessive alveolar ventilation frequently causes postoperative respiratory alkalemia, especially in hypothermic or paralyzed patients exhibiting decreased carbon dioxide production [59]. Painful attempts at arterial blood gas sampling frequently generate a spurious alkalemia.

Acute respiratory alkalemia can generate confusion or dizziness. Atrial dysrhythmias or mild cardiac conduction abnormalities sometimes appear. In patients with cerebrovascular disease, respiratory alkalemia can promote cerebral hypoperfusion and even infarction by decreasing cerebral blood flow. Reduction of serum ionized calcium ion concentration can precipitate muscle fasciculation or hypocalcemic tetany if alkalemia is severe. At very high pH levels, CNS activity, catecholamine-receptor interaction, and cardiovascular function are all severely depressed.

Metabolic compensation for acute respiratory alkalemia is limited because renal time constants for bicarbonate excretion are large. Correction of respiratory alkalemia necessitates reduction of effective alveolar ventilation, either by control of pain and anxiety using analgesics and sedatives or by reduction in mechanical ventilation. The addition of carbon dioxide to inspired gases or rebreathing of exhaled carbon dioxide have little clinical application in the PACU.

Metabolic Alkalemia

Acute metabolic alkalemia is rarely a problem in PACU patients unless alkalemia existed prior to surgery. Prolonged vomiting or gastric suctioning, pyloric stenosis, severe dehydration, excessive administration of alkaline substances, or chronic use of potassium-wasting diuretics can generate preoperative metabolic alkalemia that will persist into recovery. Large infusions of Ringer's lactate solution or blood products containing citrate can also generate postoperative alkalemia, but this usually does not appear within the first 24 hours. Occasionally, excessive intraoperative bicarbonate administration causes postoperative metabolic alkalemia.

Respiratory compensation for metabolic alkalemia through retention of carbon dioxide is rapid but somewhat limited because hypoventilation beyond a certain point causes hypoxemia. Treatment of metabolic alkalemia is directed toward resolving conditions causing the acid-base abnormality. Hydration and correction of hypochloremia and hypokalemia are important to allow correction of endogenous alkalemias. Seldom is administration of acid advisable, although intravenous hydrochloric acid infusion through a central venous catheter can be effectively utilized to treat severe, life-threatening metabolic alkalemia.

Categorization of postoperative acid-base disorders into discrete primary and compensatory couplets is sometimes difficult because rapid evolution of pathophysiology often generates two or more primary disorders. Appropriate hemodynamic, ventilatory, and metabolic support usually restores postoperative acid-base homeostasis.

■ Electrolyte and Glucose Disorders

Hypokalemia

Postoperative hypokalemia is usually accompanied by preexisting chronic potassium deficits from diuretic administration, prolonged nasogastric suctioning, vomiting, or diarrhea [62, 63]. Hypokalemia is worsened by intraoperative urinary and hemorrhagic losses, dilution during massive volume replacement, or insulin therapy. Though generally inconsequential, hypokalemia can pose a real risk of serious dysrhythmia, especially in patients taking digitalis preparations or suffering acute respiratory alkalemia. Addition of supplemental potassium to intravenous fluids is usually sufficient to keep the serum potassium concentration within an acceptable range. In selected cases, infusion of concentrated solutions through a central catheter may be necessary. Hypokalemic patients should be closely observed during infusion of calcium, insulin, or beta-sympathomimetic medications, or during periods of excessive sympathetic nervous system activity [64].

Hyperkalemia

Serious postoperative hyperkalemia occasionally occurs in patients suffering chronic renal failure or malignant hyperthermia, or after inadvertent excessive potassium administration [62]. Acute acidemia secondary to hypoventilation or hypoperfusion will exacerbate postoperative hyperkalemia [65]. Administration of succinylcholine for tracheal intubation in the PACU may increase serum potassium to dangerous levels in patients suffering from burns, multiple trauma, neurological injuries, or a neurological disease. Whenever a laboratory specimen reveals an unusually high serum potassium level with no apparent explanation, one should suspect spurious hyperkalemia caused by a hemolyzed specimen or sampling near an intravenous catheter containing potassium or banked blood. In acute postoperative settings, treatment with intravenous insulin and glucose effectively lowers the serum potassium level, and intravenous calcium temporarily counters deleterious myocardial effects. Betasympathomimetic medications may also have some value.

Hyponatremia

Postoperative hyponatremia is encountered after transurethral prostatic resection if large amounts of sodium-free irrigating solution have been absorbed by way of prostatic venous sinuses [66] or if excessive free water is infused during surgery. Symptoms of mild hyponatremia include nausea, agitation, disorientation, and visual disturbances. Severe reduction of sodium concentration causes decreased level of consciousness and hyperactive reflex activity which can progress to grand mal seizures. After prostate resection, increased serum ammonia levels caused by metabolism of absorbed glycine may exacerbate symptoms [67, 68]. Free water retention and hyponatremia can also be caused by inappropriate antidiuretic hormone (ADH) secretion, prolonged labor induction with oxytocin, excessive administration of intravenous hypotonic solutions, and respiratory uptake of nebulized droplets [69]. Therapy includes infusion of normal saline and administration of intravenous furosemide to promote renal wasting of free water in excess of sodium. In severe cases, administration of hypertonic saline may be necessary to replace a calculated sodium deficit. Careful monitoring of serum sodium concentration and osmolality is essential.

Hypocalcemia

Total body and ionized calcium concentrations can be reduced by massive fluid replacement or underlying parathyroid disease. Though often discussed, symptomatic hypocalcemia caused by extensive transfusion of blood containing chelating agents is rare, so routine administration of calcium is not usually warranted. Although major postoperative hypocalcemia seldom occurs, further reduction of the critical ionized fraction by metabolic or respiratory alkalemia may cause tetany, myocardial conduction and contractility abnormalities, and decreased vascular tone. Administration of calcium chloride in hypocalcemic patients can improve cardiovascular dynamics and response to intravenous fluid administration after major surgery.

Hyperglycemia

Elevation of serum glucose in response to stress or glucose-containing infusions is common in patients recovering from surgery [70]. Moderate hyperglycemia (200–300 mg/dl) is generally inconsequential and resolves spontaneously without treatment. It is doubtful whether transient elevation of serum glucose has any important effect on wound healing. Higher glucose levels cause glycosuria and osmotic diuresis, interfere with accuracy of serum electrolyte determinations, and increase serum osmotic load. Severe hyperglycemia can increase serum osmolality to the extent that cerebral disequilibrium and "hyperosmolar coma" supervene. In diabetic patients, hyperglycemia may also indicate severe insulin deficiency and developing diabetic ketoacidosis. Treatment includes small, incremental doses or continuous infusion of intravenous regular insulin, potassium replacement, and monitoring of serial blood glucose levels. This approach allows careful titration of blood glucose, eliminates delay in peak effect associated with longer-acting insulin preparations, and bypasses potential uptake problems with subcutaneous insulin administration. Follow-up with appropriate therapy after discharge from the PACU is critical.

Hypoglycemia

Hypoglycemia is always a serious risk, especially in the postoperative period when sedation or excessive SNS activity in recovering patients may mask signs and symptoms. Hypoglycemia in the PACU can be caused by excessive or inadvertent insulin administration or by endogenous insulin secretion. Fortunately, serious postoperative hypoglycemia is rare and easily treated by intravenous administration of 50% dextrose in conjunction with infusion of glucose-containing solutions.

Repeated accurate monitoring of serum glucose levels is important in managing postoperative hyperglycemia or hypoglycemia. Serum glucose determination is far superior to urine glucose measurement for managing blood sugar abnormalities. However, the urine glucose concentration should still be evaluated to assess osmotic diuretic effects and to help estimate renal transport of glucose by comparing urine with serum levels.

■ Hypothermia and Shivering

Many patients undergoing surgery manifest postoperative hypothermia. During general anesthesia, the temperature at which the body begins to actively regulate temperature, known as the thermoregulatory threshold, is decreased by approximately 2.5°C [71]. This allows core temperature to fall as heat is lost through radiation and convection from the skin and surgical wound, and through evaporative losses during skin preparation and respiratory humidification. Low ambient temperature in operating rooms and administration of cold intravenous fluids accelerates heat loss [72]. Once temperature falls to the thermoregulatory threshold, the only effective mechanism for heat conservation available to a paralyzed adult patient is peripheral vasconstriction. Nonshivering thermogenesis is relatively ineffective in adults, as evidenced by the minimal increase in O_2 consumption seen during hypothermia in the absence of shivering. Infants are at increased risk of intraoperative hypothermia because their body mass relative to surface area is lower. Elderly, cachectic, traumatized, or burned patients are also more prone to serious hypothermia. There is little difference in the amount or rate of heat loss with general versus regional anesthetics. However, patients receiving regional anesthetics rewarm more slowly because vasodilation and residual paralysis of striated musculature interfere with heat generation and retention.

Hypothermia can increase the risk of postoperative morbidity [73]. Hypothermic patients often exhibit increased SNS activity, elevated peripheral vascular resistance, decreased venous capacitance, and hypertension. Hypoperfusion of peripheral tissues promotes tissue hypoxia and metabolic acidemia. Viability of marginal tissue grafts can be jeopardized. Because hypothermia affects both pH and oxygen dissociation, the avidity of hemoglobin for oxygen increases, contributing to poor oxygenation of hypothermic tissues. Reduced perfusion and decreased drug biotransformation might increase the duration of action of muscle relaxants, sedatives, or hypnotics. Similarly, the minimum alveolar concentration (MAC) of inhalation anesthetics decreases approximately 5 to 7% per 1°C decrease in core temperature, accentuating sedation from residual alveolar partial pressures in the PACU. Vasoconstriction can decrease the diagnostic reliability of pulse oximetry, peripheral nerve stimulation, and intra-arterial pressure monitoring. Moderate hyperglycemia is common, and mild coagulopathy can occur secondary to visceral sequestration of platelets, decreased platelet function, and reduced activity of clotting factors. Severe hypothermia interferes with cardiac rhythm generation and impulse conduction. PR, QRS, or QT intervals can lengthen, and U waves can appear on the patient's EKG. Danger of ectopic impulse generation and dysrhythmia during mechanical stimulation of the myocardium is increased. Spontaneous ventricular fibrillation can occur if the patient's temperature falls below 28°C.

During emergence from general anesthesia, hypothalamic regulating mechanisms should increase metabolic activity and generate shivering in order to increase endogenous heat production. Severe postoperative shivering is very uncomfortable, increases the risk of incidental trauma, and makes routine postoperative care more difficult to deliver. The demand for augmented cardiac output and minute ventilation are marked, because severe shivering can increase peripheral oxygen consumption and carbon dioxide production by 200 to 300%. Myocardial ischemia or ventilatory failure can supervene in patients with coronary artery disease or limited ventilatory reserve. The intensity of postoperative shivering is sometimes accentuated by inhalation anesthetic-related tremor, which manifests both clonic and tonic components [74]. The clonic component may be related to lack of cortical influence of spinal cord reflexes, whereas the tonic component probably reflects normal shivering. Administration of morphine, meperidine, droperidol, chlorpromazine, magnesium sulfate, or methylphenidate has been advocated to suppress shivering if dangerous cardiopulmonary stress occurs [75–79]. Withholding reversal of intraoperative neuromuscular relaxants in intubated, ventilated patients can also attenuate postoperative shivering, although rewarming time is increased. However, the wisdom of administering additional relaxants during recovery just to eliminate shivering is questionable.

It is important to maintain body temperature intraoperatively by covering body and head surfaces and warming ambient air, intravenous fluids, and irrigating solutions. Use of surface or radiant warmers or heated humidification of inspired gases also minimizes postoperative hypothermia. In most cases, intraoperative core temperature falls less than 2 to 3°C and is well tolerated. A patient with an oral temperature above 35°C will spontaneously rewarm during routine recovery, although moderate shivering often occurs. Patients undergoing prolonged, major surgical procedures with extensive fluid replacement can arrive in the PACU markedly hypothermic. Temperature below 35°C is an indication for passive assistance with rewarming using radiant lighting, heating blankets, reflective coverings, or heated nebulization of inspired gas for intubated patients. All hypothermic patients should receive supplemental oxygen in recovery. As temperature rises, patients must be carefully observed for hypotension related to increasing venous capacitance and for "post-hypothermic" hyperthermia [80]. Resolution of metabolic acidemia usually accompanies rewarming, although bicarbonate administration may be required after prolonged, severe hypothermia.

■ References

1. Denlinger JK. Prolonged emergence and failure to regain consciousness. In: Orkin FK, Cooperman LH, eds. Complications in anesthesiology. Philadelphia: Lippincott, 1983:368–379
2. Longnecker DE, Grazis PA, Eggers GWN. Naloxone for antagonism of morphine induced respiratory depression. Anesth Analg 1973;52:447–453
3. Bourke DL, Rosenberg M, Allen PD. Physostigmine: effectiveness as an antagonist of respiratory depression and psychomotor effects caused by morphine or diazepam. Anesthesiology 1984;61:523–528
4. Hill GE, Stanley TH, Seutker CR. Physostigmine reversal of postoperative somnolence. Can Anaesth Soc J 1977;24:707–711
5. Jensen S, Knudsen L, Kirkegaard L, et al. Flumazenil used for antagonizing the central effects of midazolam and diazepam in outpatients. Acta Anaesthesiol Scand 1989;33:26–28
6. Ghoneim MM, Dembo JB, Block RI. Time course of antagonism of sedative and amnesic effects of diazepam by flumazenil. Anesthesiology 1989;70:899–904
7. Skillman JJ. Neurologic complications of cardiovascular surgery. I. Procedures involving the carotid arteries and abdominal aorta. In: Hindman BJ, ed. Neurological and psychological complications of surgery and anesthesia. Boston: Little, Brown, 1986:135–158
8. Shaw PJ. Neurologic complications of cardiovascular surgery. II. Procedures involving the heart and thoracic aorta. In: Hindman BJ, ed. Neurological and psychological complications of surgery and anesthesia. Boston: Little, Brown, 1986:159–174

9. Gutierrez IZ, Barone DL, Makula PA, et al. The risk of perioperative stroke in patients with asymptomatic carotid bruits undergoing peripheral vascular surgery. Am Surg 1987;53:487–489
10. Larsen SF, Zaric D, Boysen G. Postoperative cerebrovascular accidents in general surgery. Acta Anaesthesiol Scand 1988;32:698–701
11. Hindman BJ. Perioperative stroke: the noncardiac surgical patient. In: Hindman BJ, ed. Neurological and psychological complications of surgery and anesthesia. Boston: Little, Brown, 1986:101–110
12. Toung TJ, McPherson RW, Ahn H, et al. Pneumocephalus: effects of patient position on the incidence and location of aerocele after posterior fossa and upper cervical cord surgery. Anesth Analg 1986;65:65–70
13. Taenzer P, Melzack R, Jeans ME. Influence of psychological factors in postoperative pain, mood, and analgesic requirements. Pain 1986;24:331–342
14. Smiler BG, Bartholomew EG, Sivak BJ. Physostigmine reversal of scopolamine delirium in obstetric patients. Am J Obstet Gynecol 1973;116:326–329
15. Eisendrath SJ, Goldman B, Douglas J, et al. Meperidine induced delirium. Am J Psychiatry 1987;144:1062–1065
16. Hammon K, Demartino BK. Postoperative delirium secondary to atropine premedication. Anesth Progress 1985;32:107–108
17. Heath PJ, Kennedy DJ, Ogg TW, et al. Which intravenous induction agent for day surgery? A comparison of propofol, thiopentone, methohexitone, and etomidate. Anaesthesia 1988;43:365–368
18. Oliver SB, Cucchiara RF, Warner MA, et al. Unexpected focal neurologic deficit on emergence from anesthesia: a report of three cases. Anesthesiology 1987;67: 823–826
19. Dawson B, Reed WA. Anaesthesia for day care surgery: a symposium. III. Anaesthesia for surgical outpatients. Can Anaesth Soc J 1980;27:409–411
20. Ahlgren EW, Bennett EJ, Stephen CR. Outpatient pediatric anesthesiology: a case series. Anesth Analg 1971;50:402–406
21. Patel RI, Hannallah RS. Anesthetic complications following pediatric ambulatory surgery: a three-year study. Anesthesiology 1988;69:1009–1012
22. Palazzo MG, Strunin L. Anesthesia and emesis. I. Etiology. Can Anaesth Soc J 1984;31:178–187
23. Palazzo MG, Strunin L. Anesthesia and emesis. II. Prevention and management. Can Anaesth Soc J 1984;31:407–415
24. Anderson R, Crohg K. Pain as a major cause of postoperative nausea. Can Anaesth Soc J 1976;23:366–369
25. Abramowitz MD, Oh TH, Epstein BS, et al. The antiemetic effect of droperidol following outpatient strabismus surgery in children. Anesthesiology 1983;59: 579–583
26. Lerman J, Eustis S, Smith DR. Effect of droperidol pretreatment on postanesthetic vomiting in children undergoing strabismus surgery. Anesthesiology 1986;65: 322–325
27. Hardy JF, Charest J, Girouard G, et al. Nausea and vomiting after strabismus surgery in preschool children. Can Anaesth Soc J 1986;33:57–62
28. Caldamone AA, Rabinowitz R. Outpatient orchiopexy. J Urol 1982;127:286–288
29. Lindblad T, Beattie WS, Buckley DN, et al. Increased incidence of postoperative nausea and vomiting in menstruating women. Can J Anaesth 1989;36:S78–79
30. Alexander GD, Skupski JN, Brown EM. The role of nitrous oxide in postoperative nausea and vomiting. Anesth Analg 1984;63:A175
31. Lonie DS, Harper NJN. Nitrous oxide anaesthesia and vomiting. Anaesthesia 1986;41:703–707
32. Melnick BM, Johnson LS. Effects of eliminating nitrous oxide in outpatient anesthesia. Anesthesiology 1987;67:982–984

33. Muir JJ, Warner MA, Offord KP, et al. Role of nitrous oxide and other factors in postoperative nausea and vomiting: a randomized and blinded prospective study. Anesthesiology 1987;66:513–518
34. Sengupta P, Plantevin OM. Nitrous oxide and day-case laparoscopy: effects on nausea, vomiting and return to normal activity. Br J Anaesth 1988;60:570–573
35. Hovorka J, Korttila K, Erkola O. Nitrous oxide does not increase nausea and vomiting following gynaecological laparoscopy. Can J Anaesth 1989;36:145–148
36. Carter JA, Dye AM, Cooper GM. Recovery after day-case anaesthesia: the effect of different inhalational anaesthetic agents. Anaesthesia 1985;40:545–548
37. Marais ML, Maher MW, Wetchler BV, et al. Reduced demands on recovery room resources with propofol (Diprivan) compared to thiopental-isoflurane. Anesth Rev 1989;16:29–40
38. Epstein BS, Levy ML, Thein MH, et al. Evaluation of fentanyl as an adjunct to thiopental–nitrous oxide–oxygen anesthesia for short surgical procedures. Anesth Rev 1975;2:24–28
39. Hunt TM, Plantevin OM, Giblbert JR. Morbidity in gynaecological day-case surgery: a comparison of two anaesthetic techniques. Br J Anaesth 1979;51:785–787
40. Sanders RS, Sinclair ME, Sear JW. Alfentanil in short procedures: a comparison with halothane using etomidate or methohexitone for induction of anesthesia. Anaesthesia 1984;39:1202–1206
41. White PF, Chang T. Effect of narcotic premedication on the intravenous anesthetic requirement. Anesthesiology 1984;61:A389
42. Eustis S, Lerman J, Smith D. Droperidol pretreatment in children undergoing strabismus repair: the minimal effective dose. Can Anaesth Soc J 1986;33:S115
43. Wetchler BV, Collins IS, Jacob L. Antiemetic effects of droperidol on the ambulatory surgery patient. Anesth Rev 1982;9:23–26
44. Pandit SK, Kothary SP, Pandit UA, et al. Dose-response study of droperidol and metoclopramide as antiemetics for outpatient anesthesia. Anesth Analg 1989;68: 798–802
45. Poler SM, White PF. Does ephedrine decrease nausea and vomiting after outpatient anesthesia? Anesthesiology 1989;71:A995
46. Brustowicz RM, Nelson DA, Betts EK, et al. Efficacy of oral premedication for pediatric outpatient surgery. Anesthesiology 1984;60:475–477
47. Melnick BM. Extrapyramidal reactions to low-dose droperidol. Anesthesiology 1988;69:424–426
48. Cohen SE, Woods WA, Wyner J. Antiemetic efficacy of droperidol and metoclopramide. Anesthesiology 1984;60:67–69
49. Rao TLK, Madhavareddy S, Chinthagada M, et al. Metoclopramide and cimetidine to reduce gastric fluid pH and volume. Anesth Analg 1984;63:1014–1016
50. Doze VA, Shafer A, White PF. Nausea and vomiting after outpatient anesthesia: effectiveness of droperidol alone and in combination with metoclopramide. Anesth Analg 1987;66:S41
51. Gibbons PA, Nicolson SC, Betts EK, et al. Scopolamine does not prevent postoperative emesis after pediatric eye surgery. Anesthesiology 1984;61:A435
52. Loper KA, Ready LB, Dorman BH. Prophylactic transdermal scopolamine reduces nausea in postoperative patients receiving epidural morphine. Anesth Analg 1989; 68:144–146
53. Tigerstedt I, Salmela L, Aromaa U. Double-blind comparison of transdermal scopolamine, droperidol and placebo against postoperative nausea and vomiting. Acta Anaesthesiol Scand 1988;32:454–457
54. Bailey PL, Bubbers SJM, East KA, et al. Transdermal scopolamine reduces postoperative nausea and vomiting. Anesthesiology 1988;69:A641
55. Uppington J, Dunnet J, Blogg CE. Transdermal hyoscine and postoperative nausea and vomiting. Anaesthesia 1986;41:16–20

56. Askanazi J, Mordenstraum J, Rosenbaum SH, et al. Nutrition for the patient with respiratory failure: glucose vs fat. Anesthesiology 1981;54:373–377
57. Steward DJ. Malignant hyperthermia—the acute crisis. In: Britt BA, ed. Malignant hyperthermia. Int Anesthesiol Clin 1979;17:1–9
58. Grinberg R, Edelist G, Gordon A. Postoperative malignant hyperthermia episodes in patients who received "safe" anesthetics. Can Anaesth Soc J 1983;30:273–276
59. Cohen JJ, Madias NE. Acid base disorders of respiratory origin. In: Brenner BM, Stein JH, eds. Acid base and potassium homeostasis. New York, Edinburgh, London: Churchill/Livingstone, 1978:137–167
60. Relnan AS. Lactic acidosis. In: Brenner BM, Stein JH, eds. Acid base and potassium homeostasis. New York, Edinburgh, London: Churchill/Livingstone, 1978:65–100
61. Knill RL, Clement JL. Ventilatory responses to acute metabolic acidemia in humans awake, sedated, and anesthetized with halothane. Anesthesiology 1985;62:745–753
62. Kliger AS, Hayslett JB. Disorders of potassium balance. In: Brenner BM, Stein JH, eds. Acid base and potassium homeostasis. New York, Edinburgh, London: Churchill/Livingstone, 1978:168–202
63. Vitez TS, Soper LE, Wong KC, Soper P. Chronic hypokalemia and intraoperative dysrhythmias. Anesthesiology 1985;63:130–133
64. Brown MJ, Brown DC, Murphy MB. Hypokalemia from beta-2 receptor stimulation by circulating epinephrine. N Engl J Med 1983;309:1414–1419
65. Scribner BH, Fremont-Smith K, Burnell JM. The effect of acute respiratory acidosis on the internal equilibrium of potassium. J Clin Invest 1975;34:1278–1281
66. Desmond J. Complications of transurethral prostatic surgery. Can Anaesth Soc J 1970;17:25–36
67. Roesch RP, Stoelting RK, Lingeman JE. Ammonia toxicity resulting from glycine absorption during a transurethral resection of the prostate. Anesthesiology 1983; 58:577–579
68. Wang JM-L, Creel DJ, Wong KC. Transurethral resection of the prostate, serum glycine levels, and ocular evoked potentials. Anesthesiology 1989;70:36–41
69. Bancroft ML. Problems with humidifiers. In: Rendell-Baker L, ed. Problems with anesthetic and respiratory equipment. Int Anesthesiol Clin 1982;20:95–102
70. Doze VA, White PF. Effects of fluid therapy on serum glucose levels in fasted outpatients. Anesthesiology 1987;66:223–226
71. Sessler DI, Olofsson CI, Rubinstein EH, et al. The thermoregulatory threshold in humans during halothane anesthesia. Anesthesiology 1988;68:836–842
72. Morris RH, Wilkey BR. The effect of ambient temperature during surgery not involving body cavities. Anesthesiology 1970;32:102–107
73. Slotman GJ, Jed EH, Burchard KW. Adverse effects of hypothermia in postoperative patients. Am J Surg 1985;149:495–501
74. Sessler DI, Israel D, Pozos RS, et al. Spontaneous post-anesthetic tremor does not resemble thermoregulatory shivering. Anesthesiology 1988;68:843–850
75. Rodriguez JL, Weissman C, Damask MC. Morphine and postoperative rewarming in critically ill patients. Circulation 1983;68:1238–1246
76. Liem ST, Aldrette JA. Control of post anaesthetic shivering. Can Anaesth Soc J 1974;21:506–510
77. Claybon LE, Hirsch RA. Meperidine arrests postanesthetic shivering. Anesthesiology 1983;59:S180
78. Rodriguez JL, Weissman JC, Damask MC, et al. Physiologic requirements during rewarming: suppression of the shivering response. Crit Care Med 1983;11:490–497
79. Macintyre PE, Pavlin EG, Dwersteg JF. Effect of meperidine on oxygen consumption, carbon dioxide production, and respiratory gas exchange in post anesthesia shivering. Anesth Analg 1987;66:751–755
80. Martyn JW. Diagnosing and treating hypothermia. Can Med Assoc J 1981;125: 1089–1096

Special Considerations in Recovery of Children from Anesthesia

Ramesh Patel, MD
Linda Jo Rice, MD

Recovery following anesthesia and surgery is uneventful for most patients. However, for a small but significant percentage of patients, recovery from anesthesia can be a life-threatening process demanding prompt intervention by skilled nursing and medical personnel. For children undergoing anesthesia and surgery, the postanesthesia recovery period may be the most stressful and frightening time of all. Patients are anxious because they are in a new environment where they may be disoriented and are surrounded by unfamiliar people.

Two main stages of recovery are the responsibility of the anesthesiologist: care in the operating room and care in the Postanesthesia Care Unit (PACU). For patients undergoing ambulatory surgery, anesthesiologists are responsible for two more stages: recovery in the short-stay or ambulatory unit following discharge from the PACU and recovery at home following discharge from the facility.

It is essential that plans regarding the recovery process and its location be finalized before anesthesia is induced. If, for example, the patient is a former premature baby who will require monitoring for apnea postoperatively, it is prudent to make sure *before* beginning the procedure that a bed equipped to monitor both respiration and heart rate is available. Similarly, patients undergoing major thoracic, abdominal, or neurosurgical procedures who require care in special units during the recovery period must have an assured space in the appropriate unit *prior* to induction of anesthesia.

Recovery in the Operating Room

The recovery process begins in the operating room when anesthetic agents are discontinued [1]. The concentration of inhaled anesthetics is decreased, and doses of narcotics and muscle relaxants are tapered as dictated by the type of surgery and postanesthesia plans.

Pre-extubation Period

Before proceeding to tracheal extubation, the anesthesiologist should consider three things: the availability of necessary postoperative monitoring and support systems, such as an apnea monitor, a cardiac monitor, or a ventilator; and the potential need for antiemetics especially following strabismus repair or orchidopexy, or in patients with a history of motion sickness. Prophylactic antiemetics such as droperidol or metoclopramide should be considered for these high-risk patients. The third consideration is the amount of pain anticipated, plus available forms of analgesia. If the patient is expected to have little or no pain in the postoperative period, the anesthesiologist should then focus on tracheal extubation. If the patient is expected to have moderate or severe pain, however, plans to provide pain relief must be made prior to extubation, including the use of parenteral opioids. If a regional technique such as ilioinguinal nerve block, caudal block, or epidural block is preferred, then the appropriate local anesthetic block should be administered prior to extubation. Another option is to administer intravenous narcotics in the PACU.

Before proceeding with extubation, the anesthesiologist should ascertain that blood pressure, heart rate, and heart rhythm are within acceptable limits; respiratory rate and tidal volume must also be adequate. Gastric contents should be evacuated using an orogastric tube. This maneuver reduces the risk of postextubation vomiting and aspiration pneumonitis and may decrease postoperative nausea and vomiting. When the stomach contents are evacuated, the size of the stomach decreases and pressure on the lungs from increased intra-abdominal pressure is reduced; functional residual capacity (FRC) may therefore increase. The oropharynx and in some cases the tracheobronchial tree also require suctioning prior to extubation.

Extubation: Awake vs Deep

There are two options for tracheal extubation following elective general endotracheal anesthesia in an otherwise healthy patient with a normal airway: awake or deep extubation. Awake extubation entails removing the tracheal tube when the patient is awake and breathing adequately. Wakefulness is determined by return of laryngeal and pharyngeal reflexes, eye opening, grimacing, coughing, and purposeful movements. Coughing associated with awake extubation may lead to an increased incidence of postintubation croup [2].

Deep extubation is performed when the patient is still in the surgical plane of anesthesia but is breathing spontaneously and airway reflexes are depressed. Removing the tracheal tube between the awake and deep states of anesthesia leads to laryngospasm and breath holding. In a study comparing awake and deep extubation, the incidence of desaturation was equal in both groups [3]. Similarly, no difference was noted in the incidence of laryngospasm, airway obstruction, croup, or sore throat between children whose tracheas were extubated when they were awake and those whose tracheas were extubated under deep anesthesia [3]. In healthy children undergoing elective surgery, the preference of the anesthesiologist or the surgical requirement may dictate the choice of extubation technique.

Postextubation Period (Immediate)

Following extubation, oxygen is administered by mask, and the child is watched for adequacy of ventilation. Inadequate ventilation occurs due to central respiratory depression, muscle weakness, or upper airway obstruction. The most common reason for central respiratory depression during this period is medullary inhibition caused by narcotics administered during the perioperative period. Formerly premature infants are believed to be particularly sensitive to the depressant effects of narcotics. These patients should not be given narcotics and must be closely watched for central apnea following extubation. Preliminary data indicate that 10 mg/kg intravenous caffeine effectively controls apnea in otherwise healthy, formerly premature infants between 37 and 44 weeks' conceptual age who have not received perioperative narcotics [4].

The most common problem encountered following extubation is upper airway obstruction. It may be caused by enlarged adenoids or tonsils, displaced soft tissue such as the tongue (pharyngeal obstruction), or laryngospasm. Absence of breath sounds, in spite of respiratory efforts, should lead the anesthesiologist to immediately suspect upper airway obstruction. Placing the child in the lateral position or inserting an oral/nasal airway may relieve soft-tissue obstruction; however, an oral/nasal airway may irritate the patient's own airway and reflexively cause laryngospasm during lighter planes of anesthesia. Anterior jaw thrust and opening the mouth so that the tongue does not lie against the palate are generally sufficient to relieve upper airway obstruction caused by displaced soft tissue.

Laryngospasm must be differentiated from upper airway obstruction caused by soft tissues such as the tongue. Laryngospasm can quickly become life-threatening in infants and children. It is precipitated by secretions, stimulation from oral or nasal airways, or improperly timed extubation. Laryngospasm following extubation is more likely to occur in children than in adults and is rapidly followed by decreasing oxygen saturation (SpO_2). Laryngospasm (alone or in combination with some element of soft tissue obstruction) is treated by administering 100% oxygen, displacing the mandible anteriorly, opening the mouth, and applying positive pressure ventilation. If positive pressure ventilation is unsuccessful, 0.25 to 0.50 mg/kg succinylcholine is administered intravenously to relax the laryngeal musculature. Mask ventilation should then be possible. Following mask ventilation, a tracheal tube may be inserted. If a decision is made not to intubate the trachea, the child must be observed closely to ensure adequacy of ventilation. If gastric distention occurs with positive pressure ventilation via mask, then the trachea must be intubated and the distention relieved by orogastric suctioning prior to extubation.

A group that is particularly prone to upper airway obstruction is patients who have undergone tonsillectomy/adenoidectomy to relieve obstructive airway symptoms. Even after adenoids and tonsils are removed, laxity of soft palate and pharyngeal muscles may continue to cause airway obstruction. It takes 6 to 8 weeks following surgery for the pharyngeal muscles to regain full, original tone. Intraoperative narcotics are not recommended for patients with severe obstructive apnea because there is an associated central mechanism which causes apnea even after the obstruction has been relieved. Instead, narcotics should be carefully titrated in the PACU. Unfortunately, intraoperative infiltration of bupivacaine 0.25% in the peritonsillar area does not provide adequate pain relief in pediatric patients undergoing tonsillectomy [5].

Transport to the PACU

When ventilation is satisfactory, patients are transported to the PACU in the lateral position to protect against aspiration of vomitus and soft tissue–induced obstruction. Between 21 and 28% of pediatric patients exhibit arterial oxygen desaturation during transport to the PACU if supplemental oxygen is not administered [6–8]. It is therefore recommended that *all* patients recovering from general anesthesia be transported to the PACU while breathing supplemental oxygen unless adequacy of oxygenation is ensured by continuous pulse oximetry. Pediatric patients should be transported on a stretcher with raised guard rails to prevent the child from falling off. Carefully applied safety straps are also recommended.

The anesthesiologist must observe the patient carefully during the transport period. Observing chest movement and skin color, feeling for warm exhaled gases, listening to heart tones, or monitoring pulse signals are essential for all patients. Safe transport of children who have undergone major thoracic, abdominal, or cranial surgery may require a portable electrocardiogram, arterial line monitoring, and a pulse oximeter. If a decision is made to maintain tracheal intubation, a self-inflating bag with reservoir or a Mapleson-D circuit connected to an oxygen cylinder is used to ventilate the patient during transport. If the patient is being transported to an Intensive Care Unit (ICU), appropriate sizes of laryngoscopes, mask, tracheal tubes, and medications for reintubation should accompany the patient. Special attention must be paid to preventing hypothermia during transport of neonates to the nursery.

■ Postanesthesia Care Unit

The postoperative recovery period may be the most vulnerable period for children undergoing anesthesia. Nearly 50% of the perioperative cardiac arrests due to respiratory problems occur during the recovery period [9]. Neonates have the highest rate of adverse events during recovery, including cardiac arrest [10]. Children are less likely than adults to have arrhythmias or hypotension, but more likely to have airway-related problems. The overall incidence of airway problems in the PACU is 4 to 5% [10]. The PACU should be adjacent to the operating room and the ambulatory-care unit. Such an arrangement makes it easier and safer to transport patients to and from the PACU. It also allows the anesthesiologist to respond quickly in case of an emergency. The number of PACU beds required depends on the type of cases, average length of stay, and liaison with other units in the hospital. The number of nursing personnel required in a pediatric institution is higher than that in an adult hospital because of the higher turnover rate of patients in the pediatric PACU and the close observation (usually a 1-to-1 or 1-to-2 nurse-patient ratio) required in a pediatric facility.

A separate isolation room within the PACU should be available for patients with infectious diseases. Patients who are exposed to chicken pox are isolated between 10 and 28 days following exposure to protect other patients. The immunocompromised child needs special care and requires reverse isolation to protect him from airborne droplets from other children.

Common Problems in the PACU

Hypoxemia When the child arrives in the PACU, attention is focused on the cardiovascular and respiratory systems, and the temperature is determined. Oxygen should be administered to all children upon arrival in the PACU. Motoyama and Glazener [11] have shown that breathing 21% oxygen may lead to hypoxemia during the early postoperative period in children. It is recommended that an oxygen mask or blow-by oxygen be used at least until a child is awake enough to reject it, at which time the need for supplemental oxygen diminishes. Although qualitative methods of assessing oxygenation, such as observing color, may be adequate, use of pulse oximetry is encouraged and will be a requirement in the near future. The degree of wakefulness as measured by a PACU score cannot be used to establish an end point for oxygen supplementation; therefore, oxygen administration and/or SpO_2 monitoring are recommended for all children recovering from anesthesia [12]. The proposed modification of the "American Society of Anesthesiologists' Standards for Postanesthesia Care," to become effective on January 1, 1992, states that pulse oximetry *shall be* employed in children during recovery from all anesthetics during the first stage of recovery (i.e., in the PACU). Additional monitors should be used as dictated by the patient's physical condition.

Children with upper respiratory infections (URI) and those with a history of URI during the previous week should receive oxygen supplementation and/or monitoring because they are at greater risk of developing transient oxygen desaturation in the PACU [13].

Oxygen delivery is more effective if a high flow of humidified 100% oxygen is given by means of a tent or funnel device; however, it is often difficult to keep these devices attached to the child's face. Further, SpO_2 may be unreliable in an agitated child who is continuously moving.

Pain and Discomfort Postoperative pain, discomfort, and anxiety are perhaps the most common problems experienced by children in the recovery period. It is now recognized that children of all ages, including newborns and premature infants, feel pain. That preterm babies anesthetized with nitrous oxide and curare experience pain is indicated by increased levels of adrenaline, noradrenaline, glucagon, aldosterone, corticosterone, glucose, lactate, and pyruvate [14]. Intravenous fentanyl 10 μg/kg prevented this hormonal response, demonstrating that even premature infants respond to analgesia.

Management of postoperative pain starts in the operating room when intravenous narcotics or regional blocks are administered. Postoperative pain relief for children undergoing lower abdomen or lower extremity procedures can be achieved with a caudal block before the patient awakens [15]. Ilioinguinal nerve blocks, percutaneous inguinal blocks, or simple instillation of bupivacaine also provide adequate pain relief [16–18]. Adequacy of pain relief can be quantitated by using Hannallah's pain/discomfort scale (Table 1). Caudal epidural morphine has been used by Krane and co-workers [19], who noted

Table 1. *Pain/Discomfort Scale*

Observation	Criterion	Score
Blood pressure	±10% preoperative level	0
	>20% preoperative level	1
	>30% preoperative level	2
Crying	Not crying	0
	Crying but responds to tender loving care (TLC)	1
	Crying and does not respond to TLC	2
Movement	None	0
	Restless	1
	Thrashing	2
Agitation	Patient asleep or calm	0
	Mild	1
	Hysterical	2
Posture	No special posture	0
	Flexing legs and thighs	1
	Holding scrotum or groin	2
Complaints of pain (where appropriate by age)	Asleep, or states no pain	0
	Cannot localize	1
	Can localize	2
Total (maximum of 10)		

Source: Hannallah et al [16]. Used with permission.

that a dose of 0.033 to 0.1 mg/kg provided prolonged analgesia in children 1.2 to 8 years of age. Rosen and Rosen [20] found that 0.075 mg/kg morphine administered caudally was safe and effective in the treatment of postoperative pain in children following open heart surgery. However, use of caudal, epidural, or spinal narcotics is not recommended in pediatric ambulatory surgery until more information is available regarding its safety. Intravenous fentanyl 0.5 to 1.0 μg/kg or morphine 0.05 to 0.1 mg/kg are commonly used in the PACU to treat postoperative pain.

Patient-controlled analgesia (PCA) is increasingly used in pediatrics. It has been successfully used for children older than 11 years of age [21]. More recently, it has been satisfactorily used for children as young as 6 years of age [22].

Nausea and Vomiting Postoperative nausea and vomiting are relatively frequent. They often occur immediately following surgery, delaying discharge from the PACU and the ambulatory unit, and may persist even after the patient is discharged [23]. Vomiting in the PACU always carries the potential for aspiration pneumonitis in the partially obtunded patient; moreover, if attention is not paid to the volume and type of fluid infused, the child may develop hypovolemia, hyponatremia, or both. The incidence of postoperative vomiting depends on a number of factors, including age, gender, history of motion sickness, type of surgical procedure, anesthetic technique, and the skill of the anesthesiologist. The reported incidence of vomiting following strabismus surgery is as high as 80% [24]. Intraoperative use of nitrous oxide, inflation of the stomach with anesthetic gases, early postoperative ambulation, and forced feeding have all been linked to an increased incidence of postoperative emesis.

Droperidol 75 μg/kg has proven to be an effective antiemetic [24]. However, such high doses cause excessive sedation which then delays discharge [24]. Low-dose droperidol (5 μg/kg) administered before the end of the operation has been used to reduce the incidence of postoperative vomiting in children 11 to 15 years of age [25]. The routine prophylactic use of droperidol is still controversial because of the prolonged sedation associated with the effective antiemetic dose. Prophylactic intravenous lidocaine has also been used as an antiemetic [26]. Metoclopramide 0.15 mg/kg used prophylactically was found to limit emesis following strabismus surgery [27]. It did not produce sedation; as a result, patients who received metoclopramide met discharge criteria earlier than those in the control group. Patients in the control group remained in the postsurgical ambulatory unit for a longer time because of vomiting [27]. Another factor that is often overlooked is pain. It is postulated that pain may also cause nausea and vomiting. In that event, narcotics may be useful in controlling nausea and vomiting in children who are in pain [28].

Postintubation Croup The reported incidence of postintubation croup is about 1% [2]. The most common cause is a tight-fitting endotracheal tube (i.e., a tube without an air leak at 25 cm of positive pressure ventilation). Other factors associated with postintubation croup include traumatic intubation, coughing with the endotracheal tube in place, changing the patient's head position, duration of intubation, and neck surgery [2]. Children between 1 and 4 years of age are more likely to develop croup than children in any other age group. Factors that do not affect the incidence of postintubation croup are gender, presence of an upper respiratory infection (URI), use of intraoperative humidification, and analgesic lubricants. A high incidence of postintubation stridor is noted in children with Down's syndrome following cardiac surgery [29]. The incidence of croup may be decreased by administering 0.2 to 0.4 mg/kg dexamethasone prior to extubation. Any child suspected of developing croup should receive humidified mist in the recovery room. If symptoms progress or stridor develops, racemic epinephrine should be administered by aerosol. Usually 0.5 ml of 2.25% racemic epinephrine diluted in 3 to 5 ml of normal saline is aerosolized using a nebulizer and administered by inhalation with a funnel-shaped mask.

Patients should be carefully observed for recurrence of croup 2 to 4 hours following treatment with racemic epinephrine. It is our practice to admit for overnight observation any child treated with more than 1 dose of racemic epinephrine. If symptoms persist for longer than 24 hours, then other causes, such as viral croup, should be suspected. A scoring system such as Downes' Croup Scores facilitates objective evaluation and assessment of the effect of the treatment, and helps to convey the information to other medical personnel (Table 2). The use of corticosteroids in the treatment of croup is controversial.

Table 2. *Clinical Croup Score*

Sign	Score 0	1	2
Inspiratory breath sounds	Normal	Harsh with rhonchi	Delayed
Stridor	None	Inspiratory	Inspiratory and expiratory
Cough	None	Hoarse cry	Bark
Retractions and flaring	None	Flaring and supersternal retractions	As under 1 plus subcostal, intercostal retractions
Cyanosis	None	In air	In 40% O_2

Source: Downes JJ, Raphaely RC. Pediatric intensive care. Anesthesiology 1975;43:238–250. Used with permission.

Discharge from the PACU The level of nursing and medical care in the PACU is more intense than in any other section of the hospital except the Intensive Care Unit. Once discharged from the PACU, patients are not as closely monitored, so they must be carefully evaluated before discharge. Pain, nausea, and vomiting must be adequately controlled before discharge. Attention must also be paid to the patient's disposition. Although most patients can safely return to a ward, some warrant special observation. Patients who have received spinal or epidural block are not discharged until they exhibit signs of a receding block. As noted previously, formerly premature babies who are at risk for apnea, and patients with obstructive apnea (secondary to adenotonsillary hypertrophy), require a bed or space that will permit close observation and apnea monitoring.

The anesthesiologist must make sure that every patient is admitted to a bed where appropriate monitoring and nursing care are available. Patients who have undergone major abdominal or thoracic procedures may require care in the Intensive Care Unit. Patients who come to the operating room from the Intensive Care Unit or Intensive Care Nursery are generally returned to their former beds.

Discharge criteria have been developed to ensure objective evaluation of children in the PACU and to have a uniform, documented system for discharge. The Modified Aldrete Score is a 10-point system that evaluates activity, respiration, circulation (blood pressure), consciousness, and color (Table 3). It differs from the Aldrete Score, which was designed for adult patients, in three respects. The Aldrete Score evaluates dyspnea, upper- versus lower-extremity motor strength and response to verbal commands, whereas the modified scoring system focuses on the upper airway, generalized motor strength, and the ability to respond to stimuli. Steward has proposed a simplified 6-point system (Table 4). An anesthesiologist should review the patient's course and be responsible for discharging the patient from the PACU when the discharge criteria are met. Any child who does not achieve the full score should be carefully evaluated. A note documenting the evaluation and action taken should be

Table 3. *PACU Scoring System*

Parameter Observed	Score
Motor activity	
Active motion, voluntary or on command	2
Weak motion, voluntary or on command	1
No motion	0
Respiration	
Coughs on command or cries	2
Maintains good airway	1
Requires airway maintenance	0
Blood pressure	
±20 mm Hg of pre-anesthetic level	2
±20–50 mm Hg of pre-anesthetic level	1
±50 mm Hg of pre-anesthetic level	0
Consciousness	
Fully awake or easily aroused when called	2
Responds to stimuli and exhibits protective reflexes	1
No response or absence of protective reflexes	0
Color	
Pink	2
Pale, dusky, blotchy	1
Cyanotic	0
Total scores	Maximum 10

Source: Soliman et al [12]. Used with permission.

Table 4. *The Simplified Postanesthetic Recovery Score*

Observation	Score
Consciousness	
Awake	2
Responding to stimuli	1
Not responding	0
Airway	
Coughing on command or crying	2
Maintenance of good airway	1
Airway requires maintenance	0
Movement	
Moving limbs purposefully	2
Nonpurposeful movements	1
Not moving	0
Total	Maximum 6

Source: Steward DJ. A simplified scoring system for the post-operative recovery room. Can Anaesth Soc J 1975;22:111–113. Used with permission.

written prior to discharge of any patient who has not achieved the maximum score.

The anesthesiologist remains responsible for his or her patient until all risk of emergence complications has ended. At least one postoperative visit and follow-up note is required by the Joint Commission on Accreditation of Healthcare Organizations (JCAHO) in the United States. Patients who have undergone extensive procedures such as spinal fusion or cardiothoracic surgery need to be followed for a longer period. This extended contact allows us to contribute to ongoing management and evaluate the quality of our anesthetic care.

Use of patient-controlled analgesia (PCA) to treat postoperative pain in children has recently increased. The anesthesiologist typically manages pain control in these cases. Epidural and caudal catheters have also been used for postoperative pain management in children. These modalities increase the anesthesiologist's opportunity to interact as a physician with patients, parents, and other health care professionals. Anesthetic care need not end when the patient is discharged from the PACU!

■ Ambulatory Care Unit/Short-Stay Recovery Unit

Patients undergoing outpatient or ambulatory anesthesia are transferred to the short-stay recovery unit (SSRU) following discharge from the PACU. Some of the problems encountered in the PACU, such as croup, inadequate analgesia, and vomiting, also occur in the SSRU. Analysis of the reasons for admission to the hospital reveals that vomiting is the most common reason for admission (Table 5). Other common reasons are croup, fever, family request, and extensive surgery [23].

Patients are discharged home once they have met preapproved criteria (Table 6). The criteria should be rigidly enforced by the nursing staff, and any departure from them must be justified and documented.

The following discussion of criteria is a nonspecific overview of the factors that should be considered before discharge. Each institution should modify, add, or delete criteria to meet the needs of its patients, the institution, and the community.

Vital Signs

Vital signs (pulse, temperature, respiration, and blood pressure) are considered stable if they are age-appropriate and consistent with readings recorded in the operating room prior to anesthesia. It is useful to keep a table in the ambulatory unit of normal vital signs for children of various ages. All vital signs must be stable before a patient is discharged.

Gag, Swallow, and Cough Reflexes

Intact gag and swallow reflexes are necessary so that the child can drink fluids and maintain adequate hydration following discharge.

Table 5. *Reasons for Admission to the Hospital from the SSRU*[a]

Reason	No. of Patients (%)[b]
Protracted vomiting	30 (33%)
Complicated surgery	15 (17%)
Croup	8 (9%)
Parental request	6 (7%)
Fever	6 (7%)
Bleeding	3 (3%)
Sleepiness	2 (2%)
Other	20 (22%)
Total	90 (22%)

[a]Overnight hospital admission rate: 90/10,000 patients (0.9%).
[b]Percentage of those admitted.
SSRU = Short-Stay Recovery Unit.
Source: Patel et al [23]. Used with permission.

Ambulation

The child must demonstrate the ability to perform movements consistent with his or her developmental level and usual movement patterns.

Nausea, Vomiting, and Dizziness

Nausea, vomiting, and dizziness should be reduced to minimal levels before discharge. The patient should be able to sip and retain fluids. If vomiting persists, the physician should be notified.

Respiratory Distress

The patient must show no signs of respiratory distress such as stridor, retractions, nasal flaring, "barky" cough, wheezing, cyanosis, or dyspnea.

Consciousness

The patient must be alert and oriented to person, place, and time as appropriate for his or her developmental stage.

Table 6. *Discharge Criteria*

1. Vital signs ______ Stable ______ Other ______________________________
2. Reflexes ______ Swallow ______ Cough ______ Gag
3. Able to ambulate ______ Yes ______ No
4. Nausea, vomiting, dizziness ______ Minimal ______ Other ______________
5. Absence of respiratory distress ______ Yes ______ No
6. Alert and oriented ______ Yes ______ No
7. Time ______________________________
8. Date ______________________________
9. RN signature ______________________________

Source: Children's National Medical Center, Washington DC.

Voiding

Even though it is not an absolute criterion at many institutions, ability to void demonstrates return of the voiding reflex. It also helps in assessing the status of hydration of the patient.

Duration of Stay

Even if a child has met all discharge criteria, a minimum period of observation is desirable to guarantee safe discharge. At Children's National Medical Center (CNMC), patients whose tracheas were intubated during surgery are required to stay a minimum of 3 hours. Patients whose tracheas were not intubated can be discharged when the discharge criteria are met. The anesthesiologist can discharge selected patients earlier than the set limit, but only after personally examining the patient.

Fluids

Adequate hydration is a desirable goal during and after surgery. This is particularly true for procedures associated with a high incidence of vomiting, such as strabismus surgery, orchidopexy, and tonsillectomy. Patients should be given appropriate amounts of fluid before discharge, depending on their weight, their intraoperative fluid losses and fasting period.

At CNMC, when the patient meets each of the criteria it is recorded in the nurse's notes. When all criteria are met, the nurse signs the discharge sheet. Oral and written postanesthetic and postsurgical instructions are given to the parents, who then sign a release form stating that such instructions have been received. Copies of the written instructions and the signed release form are included in the patient's records.

■ Recovery Following Discharge from the Facility

Although many patients are quite active by the time they reach home following ambulatory surgery, most need an overnight rest before they can resume their usual activities. Even at home, patients continue to have minor problems and discomfort [23]. Nausea and vomiting continue to be the most common problems, as reported by parents, following discharge from the hospital (Table 7) [23].

■ Conclusion

The postoperative course of children is different from that of adults, and their care must be individualized. Vigilance and appropriate monitoring prevent many potential problems. We should make every effort to ensure that their recovery period is as pain-free and anxiety-free as possible. Children are often anxious because they are separated from their parents. Allowing parents to visit as soon as a child is awake reduces anxiety considerably. Immediate

Table 7. *Posthospitalization Complications Reported by Parents (n = 4,998)*

Complications	No. of Patients	Percentage
Vomiting (frequency)		
1–2	359	7.2%
3–4	64	1.2%
4	24	0.5%
Total (vomiting)	447	8.9%
Cough	324	6.5%
Sleepiness	297	5.9%
Sore throat	257	5.1%
Fever	235	4.7%
Hoarseness/mild croup	168	3.4%
Total	1,728	34.5%

Source: Patel et al [23]. Used with permission.

recognition and treatment of upper airway obstruction and laryngospasm are critical in the immediate postoperative period. Fluid management must be individually tailored. Attention to these details as well as "TLC" (tender loving care) will lead to a safe and comfortable recovery period.

■ References

1. Gregory GA. Pediatric anesthesia. 2nd ed. New York: Churchill/Livingstone, 1989:576
2. Koka BV, Jeon IS, Andre JM, et al. Post-intubation croup in children. Anesth Analg 1977;56:501–505
3. Patel RI, Hannallah RS, Casey WF, Verghese ST. Postoperative SpO_2 changes and complication rate following "awake" and "deep" extubation in children. Anesthesiology 1989;71:A1003
4. Welborn LG, Hannallah RS, Fink R, et al. High dose caffeine suppresses postoperative apnea in the former preterm infants. Anesthesiology 1989;71:347–349
5. Broadman LM, Patel RI, Feldman BA, et al. The effects of peritonsillar infiltration on the reduction of intraoperative blood loss and post-tonsillectomy pain in children. Laryngoscope 1989;99:578,581
6. Patel RI, Norden J, Hannallah RS. Oxygen administration prevents hypoxemia during postanesthetic transport in children. Anesthesiology 1988;69:616–618
7. Vijayakumar HR, Matriyakool K, Jewell MR. Effects of 100% oxygen and a mixture of oxygen and air on oxygen saturation in the immediate postoperative period in children. Anesth Analg 1987;66:181–184
8. Pullerits J, Burrows FA, Roy WL. Arterial desaturation in healthy children during transfer to the recovery room. Can J Anaesth 1987;34:470–473
9. Ramez Salem M, Bennett EJ, Schweiss JF, et al. Cardiac arrest related to anesthesia: contributing factors in infants and children. JAMA 1975;233:238–241
10. Cohen MM, Cameron CB, Duncan PG. Pediatric anesthesia morbidity and mortality in the perioperative period. Anesth Analg 1990;70:160–167
11. Motoyama EK, Glazener CM. Hypoxemia after general anesthesia in children. Anesth Analg 1986;65:267–272
12. Soliman IE, Patel RI, Ehrenpreis MB, Hannallah RS. Recovery scores do not correlate with postoperative hypoxemia in children. Anesth Analg 1988;67:53–56

13. DeSoto H, Patel RI, Soliman IE, Hannallah RS. Changes in oxygen saturation following general anesthesia in children with upper respiratory infectious signs and symptoms undergoing otolaryngological procedures. Anesthesiology 1988;68: 276–279
14. Anand KJS, Sippell WG, Aynsley-Green A. Randomized trial of fentanyl anesthesia in preterm babies undergoing surgery: effects on the stress response. Lancet 1987;I:243–247
15. Bramwell RGB, Bullen C, Radford P. Caudal block for postoperative analgesia in children. Anaesthesia 1982;37:1024–1028
16. Hannallah RS, Broadman LM, Belman AB, et al. Comparison of caudal and ilioinguinal/iliohypogastric nerve blocks for control of post-orchidopexy pain in pediatric ambulatory surgery. Anesthesiology 1987;66:832–835
17. Hinkle AJ. Percutaneous inguinal block for the outpatient management of post-herniorrhaphy pain in children. Anesthesiology 1987;67:411–413
18. Casey WF, Rice LJ, Hannallah RS, et al. A comparison between bupivacaine instillation versus ilioinguinal/iliohypogastric nerve block for postoperative analgesia following inguinal herniorrhaphy in children. Anesthesiology 1990;72:637–639
19. Krane EJ, Tyler DC, Jacobson LE. The dose response of caudal morphine in children. Anesthesiology 1989;71:48–52
20. Rosen KR, Rosen DA. Caudal epidural morphine for control of pain following open heart surgery in children. Anesthesiology 1989;70:418–421
21. Rogers BM, Webb CJ, Stergios D, Newman BM. Patient controlled analgesia in pediatric patients. J Pediatr Surg 1988;23:259–262
22. Dudd E, Wang JM, Rauck RL. Patient controlled analgesia for post surgical pediatric patients ages 6–16 years. Anesthesiology 1988;69:A372
23. Patel RI, Hannallah RS. Anesthetic complications following pediatric ambulatory surgery: a 3-yr study. Anesthesiology 1988;69:1009–1012
24. Abramowitz MD, Oh TM, Epstein BS, et al. The antiemetic effect of droperidol following outpatient strabismus surgery in children. Anesthesiology 1983;59: 579–583
25. Rita L, Goodarzi M, Seleny F. Effect of low dose droperidol on postoperative vomiting in children. Can Anaesth Soc J 1981;28:259–262
26. Christensen S, Farrow-Gillespie A, Lerman J. Incidence of emesis and postanesthetic recovery after strabismus surgery in children: a comparison of droperidol and lidocaine. Anesthesiology 1989;70:251–254
27. Broadman LM, Ceruzzi W, Patane PS, et al. Metoclopramide reduces the incidence of vomiting following strabismus surgery in children. Anesthesiology 1990;72: 245–248
28. Anderson R, Krogh K. Pain as a major cause of postoperative nausea. Can Anaesth Soc J 1976;23:366–369
29. Sherry KM. Postintubation stridor in Down's syndrome. Br J Anaesth 1983; 55:53–55
30. Johnson GG. Day care surgery for infants and children. Can Anaesth Soc J 1983;30:553–557
31. Steward DJ. Outpatient pediatric anesthesia. Anesthesiology 1975;43:268–276
32. Ahlgren FW, Bennett EJ, Stephen CR. Outpatient pediatric anesthesiology—a case series. Anesth Analg 1971;50:402
33. Davenport HT, Shah CP, Robinson GC. Day surgery for children. Can Med Assoc J 1971;105:498–501

Benzodiazepine Antagonists

Eran Geller, MD
Pinchas Halpern, MD

Benzodiazepines, ubiquitous in current medical practice, have an impressive safety record overall and are extensively used in anesthesia and intensive care because of their anxiolytic, sedative, and amnesic properties. The action of benzodiazepines is mediated through binding to specific receptors located on the gamma aminobutyric acid (GABA) receptor complex. By binding to these receptors, benzodiazepine agonists enhance the inhibitory effects of GABA on central nervous system (CNS) neuronal membranes, thus producing the characteristic benzodiazepine effects.

The benzodiazepine receptor is unique among CNS activating sites in that the responses to its binding with specific, different ligands can range from sedation to excitation. Thus, benzodiazepine receptor ligands may be classified as agonists, inverse agonists, or antagonists (see a previous, extensive review in this journal by Haefely [1]). The inverse agonists, by allosterically inhibiting the actions of GABA, have anxiogenic, stimulant, and convulsive effects. Antagonists, on the other hand, competitively block the actions of both agonists and inverse agonists at the benzodiazepine receptor.

A hallmark of benzodiazepines is the marked variability in response to a given dose. The extent of CNS effects and the duration of action may vary greatly from patient to patient. Factors contributing to this variability include the patient's general condition, age, and individual sensitivity, and the concomitant administration of other drugs such as opioids or barbiturates. Continuous administration or repeated, large dosing of even relatively short-acting benzodiazepines may result in prolonged effects. In spite of the relative safety of these compounds, there are a number of situations in which it may be desirable for clinicians to be able to limit or terminate the actions of benzodiazepines quickly, reliably, and safely.

The search for benzodiazepine antagonists has been a long one; drugs such as physostigmine and methylxanthines have yielded only limited success in this direction and have been associated with a high incidence of side effects. Ideally, an antagonist should be specific, lack intrinsic effects, result in rapid, complete, and reproducible antagonism of the agonist, and exhibit no important side effects. Conditions in which the availability of a benzodiazepine antagonist might be advantageous include: prolonged residual sedation beyond anesthetic needs, inadvertent oversedation, paradoxical reactions to benzodiazepines, reversal of deep sedation, and the diagnosis and treatment of intentional benzodiazepine intoxication.

The discovery of the specific antagonist flumazenil (Anexate, Hoffmann-La Roche, Basle, Switzerland) in 1979 constituted a major step forward in the search for an "ideal" benzodiazepine antagonist. Flumazenil provided, for the first time, a clear end point to benzodiazepine-induced sedation and anesthesia. Flumazenil, although at the time of this writing not yet released by the US Food and Drug Administration for clinical use in this country, is the only specific antagonist in current worldwide practice. The great amount of interest it has generated is evidenced by the fact that almost 3,000 publications have addressed its various properties in the few years since its introduction.

■ Pharmacology of Flumazenil

In 1979, an imidazobenzodiazepine was synthesized that was serendipitously found to bind to the benzodiazepine receptor with great affinity yet which, surprisingly, had almost none of the characteristic pharmacological effects of the benzodiazepines; through its binding, it prevented agonist benzodiazepine activity. Thus a specific competitive benzodiazepine receptor antagonist was discovered that was capable, through competitive mass action, of reversing the CNS effects of all benzodiazepines.

This drug flumazenil, initially named Ro 15-1788, is similar in structure to midazolam and triazolam. Both intravenous (IV) and oral formulations of flumazenil have been tested, but only the IV form is presently being marketed. Enteral absorption is rapid, but a large first-pass effect limits oral bioavailability to 16%. Distribution into the brain is very rapid, occurring within minutes and attaining brain concentrations higher than those seen in the plasma. However, flumazenil is also rapidly cleared from the brain, with a half-life of 20 to 30 minutes. The volume of distribution is approximately 1 L/kg. Onset of action is very rapid with IV administration, and an initial effect is often seen in less than 1 minute. The clinical effect and its duration are dependent on the dose of the benzodiazepine agonist and its pharmacokinetics, the dose of flumazenil, and the timing of antagonist administration. The total body elimination half-life of flumazenil is relatively short (0.7–1.3 hr) due to rapid hepatic conversion to inactive metabolites. Severe liver failure prolongs half-life to approximately 2.4 hours. Age and renal function seem to have no effect on its pharmacokinetics; clearance via hemodialysis is negligible.

The clinical duration of action is usually 1 to 3 hours but may be much shorter when large doses of benzodiazepines are combined with other CNS depressant drugs. Interestingly, very low doses of flumazenil can reverse the intense sedative and hypnotic effects of large doses of agonist. However, higher doses of flumazenil are required for antagonism of the amnesic effects of moderate doses of agonist, and still higher doses are required to reverse the anticonvulsant and anxiolytic properties of even small doses of benzodiazepines.

The currently suggested dosing of flumazenil involves administration of an initial dose of 0.1 to 0.2 mg IV, and subsequent titration with 0.1-mg increments every 30 to 60 seconds until the desired clinical effect is achieved. The titration method is employed to avoid possible anxiety reactions associated with rapid arousal. Doses of 0.2 to 0.5 mg are usually sufficient for complete antagonism of a clinical dose of benzodiazepines. Higher doses on the order of 5 to 10 mg have been administered without clinically important side effects and may be considered in individual cases. Should resedation occur, either repeat flumazenil boluses or a continuous infusion may be employed.

The present review describes the clinical use of flumazenil in anesthesia.

■ Brief Diagnostic and Therapeutic Procedures

It is the practice of many if not all physicians to provide varying degrees of sedation for short diagnostic and therapeutic procedures. Patient comfort and cooperation and the prevention of unwanted patient movement are enhanced, as is amnesia for unpleasant procedures. Residual sedation at the end of the procedure, however, is sometimes a problem even when short-acting drugs such as midazolam are employed. This concern mandates close nursing observation, necessitates transferring the patient to the recovery area, and delays discharge of the patient from the recovery area. These requirements constitute a major logistic and financial burden for medical institutions, resulting in suboptimal utilization of nursing staff and available beds and possibly detracting from overall patient safety.

The clinical efficacy and safety of flumazenil in rapidly antagonizing the sedative effects of benzodiazepines administered for sedation during brief procedures has been well established for procedures such as gastroscopy [2, 3], bronchoscopy [4], nonsurgical cardiac procedures [5], urological procedures [6], and dental procedures [7–9], among others.

In one of the initial reports describing the use of flumazenil, we studied elderly men undergoing cystoscopy and bladder biopsy under local anesthesia and midazolam hypnosis [4]. Within 7.7 ± 3 minutes following injection of a mean of 0.45 mg flumazenil, 8 of 10 patients were able to dress and walk to the holding area. Recently, Birch and co-workers [6], using a similar technique, studied in a double-blind randomized design 128 patients undergoing transurethral endoscopic procedures performed under 3 to 10 mg IM midazolam and topical anesthesia ("sedoanalgesia"). Eighty-three percent of patients receiving flumazenil were considered fit for discharge within 15 minutes, compared with 24% of the control group.

Pearson and associates [2] and Jensen and colleagues [10], among others, in double-blind placebo-controlled studies in patients undergoing gastroscopy, have also shown almost immediate, complete awakening from midazolam sedation and reversal of amnesia following flumazenil. No important side effects or resedation were observed for as long as 3 hours of observation.

In a group of 48 cardiac patients undergoing cardiac catheterization or cardioversion under midazolam or diazepam sedation, flumazenil was effective in fully reversing sedation without any hemodynamic aberration or side effects, thus permitting immediate return of the patient to the cardiac care unit at the end of the procedure [5].

Flumazenil may be especially useful following deep sedation or hypnosis with benzodiazepines administered during particularly stressful, though brief, procedures. We have successfully employed this agonist-antagonist technique for sedation during transcatheter electrical shock ablation of arrhythmogenic foci [5]. Lack of even involuntary movement during this intermittently painful procedure is crucial, yet the induction of full-scale general anesthesia is both unnecessary and fraught with risks, and would require more equipment than is normally available in the catheterization laboratory. Employing this technique, we administered IV midazolam until deep sedation was achieved and supplemented it as necessary with small doses of methohexital during defibrillation-ablation. Immediately following the procedure, deeply sedated patients were awakened by our giving incremental injections of flumazenil (up to 1 mg IV). All patients completed the procedure safely and awoke promptly following the injection of flumazenil; no complications were noted, and resedation did not occur. It is recommended that this agonist-antagonist technique be practiced only by anesthesiologists, with careful monitoring both during and after the procedure.

A number of studies have documented the safety and effectiveness of flumazenil following sedation for dental procedures. Rosenbaum and coauthors [8] administered flumazenil in a double-blind design to patients following simple dental procedures under midazolam sedation. They found effective reversal of residual sedation and anterograde amnesia. Ghoneim and associates [9], in a more detailed study, also found that a mean of 0.7 mg flumazenil was effective in reversing sedation produced by diazepam (mean of 34 mg IV) after dental surgery. However, lassitude and amnesia were only partially antagonized.

Flumazenil may thus be employed for a number of purposes in the setting of sedation for short procedures. (1) It can shorten the duration of emergence from sedation, thus allowing faster patient turnover, reducing postoperative nursing care, possibly improving safety during the recovery phase, permitting earlier patient discharge, and optimizing use of available resources [6–8]. (2) In specific procedures such as cardioversion, flumazenil can terminate deliberately deep sedation. (3) It may also be considered a "safety net" when an unintentional overdose of benzodiazepine has been administered or when an unexpectedly prolonged emergence from sedation or paradoxical reactions to benzodiazepines occur [11–12]. In this role, flumazenil is used as part of the standard resuscitation drug armamentarium and should be available wherever sedative drugs are employed.

■ Regional Anesthesia

Adequate sedation is considered by many anesthesiologists to be essential to the conduct of regional anesthesia. Properly utilized benzodiazepine sedation can improve patient comfort, alleviate anxiety, promote immobility, abolish patient recall, and possibly attenuate undesirable cardiovascular responses. Upon conclusion of the procedure, however, it may be advantageous to have an awake and cooperative patient. Occasionally, inadvertently deep sedation is produced, especially with elderly or debilitated patients in whom it may be difficult to accurately titrate sedative drugs.

Halim and co-workers [13] recently reviewed the many studies that have documented the safe and effective use of flumazenil in awakening patients sedated with diazepam, flunitrazepam, or midazolam during regional anesthesia. Reversal of sedation and amnesia, coupled with a return to presedation levels of orientation, comprehension, and collaboration, was reported with flumazenil in doses ranging from 0.005 to 0.1 mg/kg. Full antagonism was always evident by 5 minutes, and no resedation was reported. Side effects occasionally occurred after flumazenil administration but were not serious or prolonged, nor did they require treatment. In early studies, anxiety associated with awakening was often reported, attributed to the rather rapid injection of large doses (0.1 mg/kg) in the original tests. In more recent investigations, when flumazenil was given by titration to a maximum dose of 1 mg, no awakening anxiety was observed.

It is doubtful that all sedated patients would benefit from flumazenil following regional anesthesia, but it is reassuring to have the antagonist available for patients who are unusually deeply sedated or in whom recovery is prolonged.

■ General Anesthesia

Flumazenil has been shown to be highly effective and safe in reversing the residual effects of benzodiazepines administered during general anesthesia. In several studies of anesthesia for short gynecological procedures such as dilatation and curettage or laparoscopy [14–17], flumazenil in doses of 0.005 to 0.10 mg/kg was found to produce rapid and complete awakening from benzodiazepine-based general anesthesia. No important side effects were observed. When general anesthesia for major surgical procedures was studied, the results were similar. Thus, Skielboe and associates [18] found in a double-blind placebo-controlled study that flumazenil (median dose 0.35 mg) was fully effective in rousing surgical patients after general anesthesia during which a mean of 0.33 mg/kg diazepam had been administered. No side effects or resedation were noted during the 4-hour observation period. Zuurmond and coauthors [19] studied in a randomized fashion 40 day-case surgical patients anesthetized with midazolam and isoflurane, and found that flumazenil 0.5 ± 0.2 mg hastened recovery and had no side effects.

Flumazenil seems to be a safe, effective drug to use to counteract excessive residual sedation following general anesthesia that includes benzodiazepines. It permits rapid assessment of a patient's CNS status in the situation of delayed emergence and thus may improve the safety and ease of postoperative care. Complete reversal of neuromuscular blockade should always be ensured before antagonism of sedation is attempted, so that tracheal extubation may be effected as soon as the patient shows signs of awakening following the administration of flumazenil.

■ Total Intravenous Anesthesia

Total intravenous anesthesia (TIVA), of which benzodiazepines are mainstays, is gaining in popularity. One of the potential difficulties associated with TIVA is residual sedation [20, 21]. Prolonged recovery is not unexpected, given the interaction between the rather large doses of benzodiazepines and the opioids commonly used in this technique. The use of flumazenil to enhance recovery and minimize the risks from prolonged postoperative somnolence seems logical. Indeed, a number of studies have documented the safety and efficacy of flumazenil in this setting [16, 20–22]. Hennis and colleagues [22], for example, in a randomized study, showed full clinical reversal of deep residual sedation following high doses of midazolam (approximately 1 mg/kg). Psychomotor tests, however, showed residual impairment, and evidence of resedation was noted after approximately 60 minutes.

Although flumazenil may be used successfully following TIVA, care must be exercised that patients not be discharged prematurely from the recovery room. Probably at least two half-life times of flumazenil (i.e., 1.5–2.5 hr) should elapse before a patient is presumed safe from the development of resedation. Consideration in this situation should be given to the possibility of administering repeat doses of flumazenil or to instituting a continuous infusion in order to prevent potential complications.

■ Effects in Elderly and Very Ill Patients

Ricou and co-workers [11] and Sage and associates [23] showed flumazenil to be successful in reversing benzodiazepine sedation for regional anesthesia in 62 elderly patients. The only notable side effect was a 10% incidence of anxiety in a subgroup of 30 patients who were reversed with a rapid and large dose of flumazenil (0.1 mg/kg); even so, no hemodynamic aberrations were observed. Juettner and coauthors [24] studied very ill cancer patients undergoing intraluminal "afterloading" radiation therapy under heavy midazolam sedation (0.12 mg/kg/hour infusion plus boluses). In patients remaining deeply sedated, a flumazenil 0.5-mg IV bolus followed by a continuous infusion of 0.15 μg/kg/min over 30 to 60 minutes resulted in immediate and sustained reversal of sedation. Most of the patients, however, were described as being anxious on awakening. Fragen and colleagues [25] recently studied a group of elderly patients sedated with midazolam during regional anesthesia, in whom flumazenil in a dose of 0.4 to 0.6 mg was sufficient to antagonize sedation without causing any important side effects. It may thus be concluded that flumazenil is safe and effective even in elderly and debilitated patients, who are prone to become overly sedated even with very small doses of benzodiazepines.

■ Electroencephalographic Effects

Flumazenil administration to patients under the influence of benzodiazepines produces rapid reversal of benzodiazepine-induced changes in the electroencephalogram. Crozier and co-workers [26] demonstrated an immediate increase in alpha and beta frequency power after flumazenil. Freye and associates [27] also found a decrease in the power of the slow delta band and an increase in the beta range in intensive care patients receiving benzodiazepine sedation. Kochs and colleagues [28] studied auditory brain stem evoked responses and found a reversal of benzodiazepine-induced prolongation of latencies.

■ Hemodynamic and Respiratory Effects

The hemodynamic effects associated with flumazenil administration have been studied both in healthy patients and in patients with cardiac disease; hemodynamic stability is the rule. No appreciable cardiovascular effects were noted after reversal of benzodiazepine-induced sedation during cardiac catheterization [29–33] in patients with severe cardiac disease or following cardiac surgery [32]. Almost all studies of flumazenil, even in patients with intentional drug overdose, have noted stable hemodynamics following antagonism of benzodiazepine-induced sedation. These findings seem somewhat surprising at first, considering that patients awaken fairly abruptly, often into unfamiliar and rather intimidating surroundings, but cardiovascular changes generally do not occur.

The lack of hemodynamic effects may be at least partially explained by findings showing that antagonism of the CNS effects of benzodiazepines by flumazenil induces only minimal changes in the acute stress hormones such as epinephrine, norepinephrine, vasopressin, and beta endorphin [31, 32].

Studies regarding the effects of flumazenil on respiratory depression caused by benzodiazepines have produced somewhat conflicting results. Although flumazenil transiently antagonized the hypoxemia induced by midazolam in volunteers, it may also have enhanced hypoxemia caused by fentanyl [34, 35], a finding that suggests an intrinsic agonistic effect of flumazenil at medullary respiratory centers. In another study, the depression of the isocapneic hypoxic ventilatory drive by a very large dose of diazepam (1.6 ± 0.8 mg/kg) was not reliably antagonized by a dose of flumazenil sufficient to fully reverse clinical sedation [36]. Weinbroum and Geller [37] however, studied patients recovering from surgery under regional anesthesia and sedated with midazolam alone or with narcotics, and patients undergoing general anesthesia (unpublished data, 1989). These investigations showed that reversal of sedation with flumazenil antagonized the slight fall in arterial oxygen saturation caused by sedation with benzodiazepines alone, and had no deleterious effect on arterial saturation in patients sedated or anesthetized with midazolam and opioids.

The depression of respiratory rate induced by midazolam was fully reversed by flumazenil in a study by Ritter and Hoshikazi [38]. In a study by Carter and colleagues [39], however, a mean dose of 6 mg midazolam resulted in a significant fall in respiratory minute volume, which did not revert immediately to baseline, even when patients became fully awake within 1 to 2 minutes following as much as 1 mg flumazenil. In conclusion, presently available data are equivocal as to whether flumazenil is effective in fully reversing the respiratory effects of the benzodiazepines. Therefore, it should not be taken for granted that a patient fully awakened by flumazenil is also free of residual respiratory depression.

Effects on Intracranial Pressure and Cerebral Hemodynamics

The effects of flumazenil on intracranial pressure (ICP) and cerebral hemodynamics have not been studied very extensively in the clinical setting. Forster and co-workers [40] have shown in healthy volunteers that flumazenil (0.1 mg/kg) has no intrinsic effects on cerebral blood flow (CBF). Flumazenil did, however, prevent the 30% reduction in CBF caused by unopposed midazolam. Fleischer and associates [41] also showed (in dogs) that flumazenil in a large dose (1.0 mg/kg) returned the cerebral metabolic rate for oxygen to baseline following its depression by large doses of midazolam. In goats subjected to incomplete brain ischemia, antagonism of midazolam by flumazenil was associated with significant increases in CBF and ICP, both of which may have been caused by an increase in arterial blood pressure in the presence of impaired cerebrovascular autoregulation [42]. In a study of patients with severe head injury sedated with midazolam, administration of flumazenil (0.1-mg increments, up to 1 mg IV) caused a clinically significant ICP increase in patients where ICP was poorly controlled (the mean increasing from 16.3 to 25.2 mm Hg) [43]. No such change occurred, however, in a subgroup of patients with well-controlled ICP. The use of flumazenil at the end of neurosurgical procedures, where rapid awakening is desirable in order to allow accurate neurological assessment, has, to our knowledge, been reported only once. Chiolero and co-investigators [44] in an open, uncontrolled study of 18 patients found that flumazenil up to 1 mg resulted in rapid awakening after neurosurgical procedures for which a midazolam infusion constituted part of the anesthetic. Resedation in some patients and transient hypertension in others were noted. At the present time, and pending further studies, it seems that flumazenil reversal of benzodiazepine effects should be employed cautiously, if at all, in patients with severe brain damage, especially if increased ICP and poor intracranial compliance are known or presumed to exist.

Intrinsic Effects of Flumazenil

Flumazenil was initially assumed to be a "pure" antagonist, devoid of intrinsic effects. Later, evidence accumulated showing that flumazenil possesses mild intrinsic agonistic as well as inverse agonistic effects. For example, flumazenil exhibits an anticonvulsant (agonistic) effect that is currently being evaluated in the treatment of epilepsy [45, 46]. Other studies found that flumazenil could increase anxiety, which implies a mild inverse agonistic effect [44]. In a study by Higgitt and colleagues [47], oral flumazenil 30 to 100 mg exhibited a mild agonistic activity on physiological parameters (slight decrease in arterial blood pressure and respiratory rate) but simultaneously, an inverse agonistic or anxiogenic activity on subjective mood measures such as contentedness. With the postulation of the existence of endogenous benzodiazepine receptor ligands (both agonists and inverse agonists), the search for "intrinsic" effects of the benzodiazepine antagonists in general, and flumazenil in particular, has acquired new meaning and intensity. There is evidence, for example, that flumazenil may reverse the coma associated with hepatic encephalopathy [48], a condition in which endogenous benzodiazepine ligands have been postulated to exist. The clinical significance of flumazenil's intrinsic effects is unclear at the present time, and conclusions must await further studies.

■ Flumazenil and Benzodiazepine Withdrawal Effects

Flumazenil may precipitate acute withdrawal symptoms in subjects physically dependent on benzodiazepines [1]. Most data suggesting this relationship, however, were derived from animal studies, and their applicability to the clinical setting, especially with the currently recommended low doses, is far from clear. Danton and co-workers [49] found no clinical signs of withdrawal or seizures in groups of volunteers given flumazenil who had ingested 15 mg of diazepam per day for 14 days. Also, the extensive reported experience with antagonism of intentional benzodiazepine overdoses by flumazenil in hundreds of patients, many of them chronic benzodiazepine users [50, 51], shows an impressive safety record. Thus, it seems at the present time that the risk of precipitating symptoms of withdrawal from chronic benzodiazepine use by flumazenil is not great.

■ Residual Psychomotor Effects and Resedation

Mild to moderate residual psychomotor impairment after clinically complete antagonism of sedation with flumazenil has been noted by some investigators. Although following flumazenil administration, patients seem to be fully alert as judged by sedation scoring, more in-depth testing sometimes reveals fine motor and cognitive deficits. Short and Galletly [52] have shown in volunteers that residual psychomotor impairment persisted following antagonism of 10 mg of midazolam with 1 mg or 4 mg of flumazenil. The authors concluded that the usually recommended flumazenil dose of 1 mg may be insufficient to fully antagonize a recently administered, relatively large dose of benzodiazepines. The larger dose of flumazenil had a more complete antagonizing effect on most variables tested. Gentil and coauthors [53] also noted that a mean of 5.3 mg flumazenil was required for full reversal of the psychomotor effects of 2 mg of flunitrazepam administered to volunteers. The import of residual psychomotor deficits remains unclear and is probably minor. However, even fully awake patients should be considered to have residual psychomotor impairment and therefore warned not to drive or operate machinery for up to 24 hours.

Resedation following antagonism of benzodiazepines with flumazenil is due to the shorter half-life of the antagonist, as compared with that of most clinically used benzodiazepines. In a recent review of the literature regarding this topic, Whitwam [54] concluded that in most instances where clinically appropriate doses of benzodiazepines were used, resedation was not a problem. Considerable resedation has been reported, however, following large doses of benzodiazepines given for hypnosis as part of total intravenous anesthesia, or ingested as an intentional benzodiazepine overdose. Renewed sedation or hypnosis may become manifest after a period of between 15 minutes and 2 hours. Resedation may ensue even after clinically complete antagonism of benzodiazepine effects, and even with relatively short-acting drugs such as midazolam given in large doses.

The likelihood of resedation, its severity, and its time of appearance probably depend on the dose and the half-life of the benzodiazepine employed, as well as on the timing and the dose of flumazenil administered [54]. The general condition of the patient, individual variability, and concomitant administration of drugs such as opioids or other CNS depressants also determine to a large extent the patient's initial and subsequent responses to the administered benzodiazepine.

Resedation may be a major hazard to the patient, requiring continued close observation and sometimes repeat bolus doses of flumazenil or the administration of a continuous infusion. It has been our experience in treating patients with benzodiazepine overdose [55] that some patients need multiple repeat boluses or continuous administration of 0.1 to 0.5 mg/hr of flumazenil until the benzodiazepine has been metabolized, and sedation does not recur. All such patients should be observed under strict PACU protocol until they have shown no rebound of benzodiazepine effects for at least 2½ hours.

■ Reversal of the Effects of Inhalation Anesthetics

It has recently been suggested that the GABAergic system may be involved in the mechanism of action of inhalation anesthetics. Specifically, Geller and associates [56] showed that flumazenil increased motor activity of mice recovering from halothane anesthesia, although the minimum alveolar concentration (MAC) of halothane in mice and rats was not influenced by flumazenil [56, 57]. In a double-blind study of 96 patients undergoing elective surgery under mask halothane anesthesia without benzodiazepines, flumazenil was found to improve the quality of emergence from anesthesia [58]. Recently, partial inverse agonists of the benzodiazepines were shown to significantly shorten recovery of rats from methoxyflurane anesthesia [59]. Evidence that these drugs affect inhalation anesthetics holds promise for developing a new pharmacological modality to enhance recovery from anesthetic agents.

■ Conclusion

Flumazenil has proven a highly effective and specific benzodiazepine antagonist. It provides a safe means of rapidly attenuating or reversing the CNS-depressant effects of benzodiazepines whenever indicated. The availability of flumazenil has led to the development of an anesthetic technique based on the agonist-antagonist concept, which allows accurate termination at will of benzodiazepine-induced sedation or anesthesia. Utilization of this process may improve the management of day-case anesthesia, recovery room procedures, and sedation in critically ill ventilated patients, and may enhance diagnosis and treatment of life-threatening intoxications in the emergency department. Under no circumstances, however, should flumazenil be relied upon to the extent that excessive doses of benzodiazepines become used complacently. Even an eminently safe antagonist like flumazenil should not serve as a substitute for the correct and careful use of anesthetics. As further experience with flumazenil is gained and it becomes more widely available, other applications may be forthcoming.

■ References

1. Haefely W. Benzodiazepines. Int Anesthesiol Clin 1988;26:262–272
2. Pearson RC, McCloy RF, Morris P, Bardhan KD. Midazolam and flumazenil in gastroenterology. Acta Anaesthesiol Scand 1990;34(suppl 92):21–24
3. Sanders LD, Davies-Evans J, Rosen M, Robinson JO. Comparison of diazepam with midazolam as IV sedation for outpatient gastroscopy. Br J Anaesth 1989;63: 726–731
4. Geller E, Silbiger A, Niv D, et al. The reversal of benzodiazepine sedation with Ro 15-1788 in brief procedures. Anesthesiology 1986;65:A357
5. Geller E, Halpern P, Silbiger A, et al. The use of Anexate (Ro 15-1788) for the reversal of benzodiazepine sedation in cardiac patients. Proceedings of the World Federation of Anesthesiologists Association Congress, Washington DC, 1988
6. Birch BRP, Anson K, Gelister J, et al. The role of midazolam and flumazenil in urology. Acta Anaesthesiol Scand 1990;34(suppl 92):25–32
7. Young ER, Kestenberg SH, Toal CB. The effect of flumazenil on the recovery time of dental patients sedated with diazepam. Anesth Prog 1989;36:46–51
8. Rosenbaum NL, Hooper PA. The effects of flumazenil, a new benzodiazepine antagonist, on the reversal of midazolam sedation and amnesia in dental patients. Br Dent J 1988;165:400–403
9. Ghoneim MM, Dembo JB, Block RI. Time course of antagonism of sedative and amnesic effects of diazepam by flumazenil. Anesthesiology 1989;70:899–904
10. Jensen S, Knudsen L, Kirkegaard L, et al. Flumazenil used for antagonizing the central effects of midazolam and diazepam in outpatients. Acta Anaesthesiol Scand 1989;33:26–28
11. Ricou B, Forster A, Bruckner A, et al. Clinical evaluation of a specific benzodiazepine antagonist (Ro 15-1788): studies in elderly patients after regional anaesthesia under benzodiazepine sedation. Br J Anaesth 1986;58:1005–1011
12. Stolarek IH, Ford MJ. Acute dystonia induced by midazolam and abolished by flumazenil. Br Med J 1990;300:614
13. Halim B, Schneider I, Claeys MA, Camu F. The use of midazolam and flumazenil in locoregional anaesthesia: an overview. Acta Anaesthesiol Scand 1990;34(suppl 92):42–46
14. Alon E, Baitella L, Hossli G. Double blind study of midazolam supplemented general anesthesia with Ro 15-1788. Br J Anaesth 1987;49:455–458
15. Hamar O, Garamvolgyi Gy, Kalman A. Fentanyl-midazolam-flumazenil anesthesia for induced abortion. Int J Gynaecol Obstet 1989;30:69–72
16. Raeder JC, Nilsen OG, Hole A. Pharmacokinetics of midazolam and alfentanil in outpatient general anesthesia: a study with concomitant thiopentone, flumazenil or placebo administration. Acta Anaesthesiol Scand 1988;32:467–472
17. Rupp SM, Stephenson CA. Flumazenil versus placebo in antagonism of midazolam-based general anesthesia for outpatient surgery. Anesth Analg 1990;70(suppl 2):S340

18. Skielboe M, Andersen P, Weber M, et al. Antagonism of diazepam sedation by flumazenil. Br J Anaesth 1989;63:554–557
19. Zuurmond WWA, van Leeuwen L, Helmers JHJH. Recovery from fixed-dose midazolam-induced anaesthesia and antagonism with flumazenil for outpatient arthroscopy. Acta Anaesthesiol Scand 1989;33:160–163
20. Klaussen NO, Juhl O, Sorensen AH, et al. Flumazenil in total intravenous anaesthesia using midazolam and fentanyl. Acta Anaesthesiol Scand 1988;32:409–412
21. Nilsson A, Persson MP, Hartvig P. Effects of the benzodiazepine antagonist flumazenil on postoperative performance following total intravenous anaesthesia with midazolam and alfentanil. Acta Anaesthesiol Scand 1988;32:441–446
22. Hennis PJ, van Haastert FA, Mulder AJ, Spierdijk J. Antagonism of midazolam sedation by flumazenil: a placebo-controlled study in patients recovering from intravenous anaesthesia with high doses of midazolam. Eur J Anaesthesiol 1988;5:369–376
23. Sage DJ, Close A, Boas RA. Reversal of midazolam sedation with Anexate. Br J Anaesth 1987;59:459–464
24. Juettner FM, Ratzenhofer B, Friehs G, Hackl A. Short term anaesthesia with midazolam and flumazenil during intraluminal afterloading of esophageal and bronchial neoplasms in ill, inoperable patients. Endoscopy 1990;22:97–98
25. Fragen RJ, Katz JA, Dunn KL. Flumazenil reversal of midazolam sedation in the elderly. Anesth Analg 1990;70(suppl):S113
26. Crozier THA, Sydow M, Muller J, et al. Hemodynamic effects of the benzodiazepine antagonist flumazenil after laparotomies under total intravenous anesthesia. Anaesthesist 1989;38:193–199
27. Freye E, Neruda B, Falke K. Flumazenil (Anexate[R]) for the reversal of residual benzodiazepine activity. Drugs of today. 1989;25:119–124
28. Kochs E, Wuest P, Schulte am Esch J. Acoustic and somatosensory evoked potentials under sedation with midazolam and antagonizing via flumazenil. Anasth Intensivther Notfallmed 1989;24:49–56
29. Geller E, Halpern P, Chernilas Y, Miller HI. Cardiorespiratory effects of flumazenil in cardiac patients. Anesth Analg (in press)
30. Marty J, Joyon D. Haemodynamic responses following reversal of benzodiazepine-induced anaesthesia or sedation with flumazenil. Eur J Anaesthesiol 1988;suppl 2:167–171
31. Croughwell ND, Reves JG, Will CJ, et al. Safety of flumazenil in patients with ischaemic heart disease. Eur J Anaesthesiol 1988;suppl 2:177–182
32. Croughwell ND, Reves JG, Will CJ, et al. Safety of rapidly administered flumazenil in patients with ischemic heart disease. Anesth Analg 1990;70:S73
33. Louis M, Forster A, Suter PM, Gemperle M. Clinical and hemodynamic effects of a specific benzodiazepine antagonist (Ro 15-1788) after open heart surgery. Anesthesiology 1984;61:A61
34. Tolksdorf W, Prag HO, Amberger M, Vorwold M. The respiratory depression caused by fentanyl is slightly enhanced by flumazenil. Anesth Analg 1990;70:S410
35. Tolksdorf W, Ney Ch, Ney R, Amberger M. The influence of flumazenil on respiration after midazolam and/or fentanyl. Anesth Analg 1990;70:S409
36. Mora CT, Torjman M, White PF. Effects of diazepam and flumazenil on sedation and hypoxic ventilatory response. Anesth Analg 1989;68:473–478
37. Weinbroum A, Geller E. The respiratory effects of reversing midazolam sedation with flumazenil in the presence or absence of narcotics. Acta Anaesthesiol Scand 1990;34(suppl 92):65–69
38. Ritter JW, Hoshikazi D. Flumazenil antagonizes midazolam-induced respiratory rate depression. Anesth Analg 1990;70(suppl):S324
39. Carter AS, Bell GD, Coady T, et al. Speed of reversal of midazolam-induced respira-

tory depression by flumazenil: a study in patients undergoing upper GI endoscopy. Acta Anaesthesiol Scand 1990;34(suppl 92):59–64
40. Forster A, Juge O, Louis M, Nahory A. Effects of a specific benzodiazepine antagonist (Ro 15-1788) on cerebral blood flow. Anesth Analg 1987;66:309–313
41. Fleischer J, Milde J, Michenfelder D. Cerebral effects of high dose midazolam and subsequent reversal with Ro 15-1788 in dogs. Anesthesiology 1988;68:234–242
42. Kochs E, Roewer N, Schulte am Esch J. Effects of flumazenil on cerebral blood flow and intracranial pressure after incomplete cerebral ischemia in goats. Anesthesiology 1988;69:A586
43. Chiolero RL, Ravussin P, Anderes JP, et al. The effects of midazolam reversal by Ro 15-1788 on cerebral perfusion pressure in patients with severe head injury. Intensive Care Med 1988;14:196–200
44. Chiolero RL, Ravussin P, Chassot PG, et al. Ro 15-1788 for rapid recovery after craniotomy. Anesthesiology 1986;65:A466
45. Klotz U, Kanto J. Pharmacokinetics and clinical use of flumazenil (Ro 15-1788). Clin Pharmacokinet 1988;14:1–12
46. Scollo-Lavizzari G. The clinical anticonvulsant effects of flumazenil, a benzodiazepine antagonist. Eur J Anaesthesiol 1988;suppl 2:129–138
47. Higgitt D, Lader M, Fonagy P. The effects of the benzodiazepine antagonist Ro 15-1788 on psychophysiological performance and subjective measures in normal subjects. Psychopharmacology 1986;89:395–403
48. Jones EA, Skohick P, Gammal SH, et al. The gamma aminobutyric acid A ($GABA_A$) receptor complex and hepatic encephalopathy: some recent advances. Ann Intern Med 1989;110:532–546
49. Danton AW, Schwam E, Pitman V, et al. Flumazenil: US clinical pharmacology studies. Eur J Anaesthesiol 1988;suppl 2:81–95
50. Lheureux P, Askenasi R. Agonism and antagonism at the benzodiazepine receptor: a new concept for new approach of common drug poisoning. Acta Anaesthesiol Belg 1989;40:23–28
51. Weinbroum A, Halpern P, Geller E. The use of flumazenil in drug overdoses. Intensive Care Med (in press)
52. Short TG, Galletly DC. Residual psychomotor effects following reversal of midazolam sedation with flumazenil. Anaesth Intensive Care 1989;17:290–297
53. Gentil V, Tavares S, Gorenstein C, et al. Acute reversal of flunitrazepam effects by Ro 15-1788 and Ro 15-3505: inverse agonism, tolerance and rebound. Psychopharmacology 1990;100:54–59
54. Whitwam JG. Resedation. Acta Anaesthesiol Scand 1990;34(suppl 92):70–74
55. Geller E, Niv D, Weinbroum A, et al. The use of flumazenil in the treatment of 34 intoxicated patients. Resuscitation 1988;suppl 16:57–62
56. Geller E, Schiff B, Halpern P, et al. A benzodiazepine receptor antagonist improves emergence of mice from halothane anesthesia. Neuropharmacology 1989;28: 271–274
57. Greiner AS, Larach DR. The effect of benzodiazepine receptor antagonism by flumazenil on the MAC of halothane in the rat. Anesthesiology 1989;70:644–648
58. Geller E, Weinbroum A, Halpern P, et al. Flumazenil, the specific benzodiazepine receptor antagonist (Ro 15-1788), affects recovery from halothane anesthesia. Anesthesiology 1987;67:A422
59. Miller DW, Yourick DL, Tessel RE. Antagonism of methoxyflurane induced anesthesia in rats by benzodiazepine inverse agonists. Eur J Pharmacol 1989;173:1–10

Opioid Antagonists

Paul H. Alfillé, MD
Carl E. Rosow, MD, PhD

Opioid analgesics, still central to the practice of modern anesthesia, are today used as primary or adjuvant anesthetics, sedatives, and postoperative analgesics. Opioid analgesia comes at a price, however; clinically useful concentrations produce a constellation of depressant and stimulant side effects. A large part of recovery-room care is concerned with treating the effects of too much or too little opioid.

Opioids may be divided into three broad classes based on their pharmacodynamic effects: pure agonists (morphine, meperidine, fentanyl), pure antagonists (naloxone, naltrexone), and mixed agonist-antagonists (nalbuphine, butorphanol, buprenorphine, pentazocine). This review will focus on the clinical pharmacology of the last two of these classes—opioids with antagonist properties. This heterogeneous group of drugs possesses a broad range of clinical applications:

1. In the recovery room, the emergency ward, or the delivery suite, the primary application for a pure antagonist is reversal of respiratory depression caused by pure agonists like morphine or fentanyl.
2. It has now become common to use pure or mixed antagonists to reverse inconvenient or uncomfortable opioid effects like pruritis and urinary retention. Antagonists are particularly useful when the analgesic has been administered epidurally or intrathecally.
3. Mixed agonist-antagonists have been advocated to reverse respiratory depression while preserving postoperative analgesia.
4. Mixed agonist-antagonists are frequently used as intraoperative and postoperative analgesics or sedatives in their own right.

Antagonists and Opiate Receptors

The prototypical opioid antagonist is nalorphine (N-allylnormorphine), discovered to be a morphine reversal agent prior to World War II. The initial animal studies did not indicate that this compound had any analgesic effects.

Table 1. *Opioid Receptors and Their Physiological Effects*

Effect	Receptor		
	Mu	Kappa	Sigma
Analgesia	Yes	Yes	No
Respiration	Depression	Depression	Stimulation
Mood	Euphoria	Sedation	Dysphoria
Pupil	Miosis	Miosis	Mydriasis
Morphine withdrawal	Suppression	No effect	No effect

Source: Martin et al [3].

Lasagna and Beecher [1] studied nalorphine in postoperative patients and were surprised to discover that it was nearly as potent an analgesic as morphine. Unfortunately, the drug also turned out to be a powerful hallucinogen, and few patients would agree to take it a second time.

Wikler and colleagues [2] made the important discovery that nalorphine would not sustain opioid-type physical dependence. He administered nalorphine to morphine-dependent subjects and showed that it produced a particularly violent "precipitated withdrawal" syndrome. Although its psychotomimetic properties made it unusable as an analgesic, nalorphine was introduced into clinical practice as Nalline, the first narcotic antagonist. This drug also generated intense interest in the pharmaceutical industry because it was the first potent opioid analgesic that had little or no addiction liability.

How can a single drug be both an agonist (analgesic) and an antagonist at the same time? The simplest explanation was formulated by William Martin, who postulated the existence of two different opioid receptors which he later named mu and kappa [3]. Morphine, he reasoned, has affinity and efficacy at mu receptors; nalorphine produces its effects at kappa receptors. Nalorphine, in addition, binds to mu receptors, where it acts as a competitive antagonist. The pure antagonist, naloxone, reverses both morphine and nalorphine because it binds to both receptors but does not have efficacy at either one. Pentazocine, butorphanol, and nalbuphine are newer agonist-antagonist opioids that also appear to act at kappa receptors. Mu and kappa receptors, unfortunately, are not as well characterized as autonomic receptors, but we believe that they mediate distinct sets of biological effects (Table 1). Martin also proposed a sigma receptor to account for the psychotomimetic properties of nalorphine, but it is now uncertain whether sigma receptors are specific for opioids.

Buprenorphine is an analgesic thought to produce both its agonist and antagonist effects at a single opioid receptor. It binds with high affinity to mu receptors but has less peak analgesic effect than a pure agonist like morphine or fentanyl. Buprenorphine is therefore called a mu partial agonist. When buprenorphine is given after a large dose of morphine, it displaces the pure agonist and reduces the intensity of both analgesia and respiratory depression, as well as of other agonist actions.

Each of the drugs we will discuss can be characterized as an agonist, partial agonist, or antagonist at mu, kappa, or sigma receptors (Table 2). These drugs vary widely in their potency as antagonists (Table 3).

It is important to remember that both the desirable and the undesirable

Table 2. *Receptor Selectivity of Opioid Agents*

Drug	Receptor		
	Mu	Kappa	Sigma
Morphine	Agonist	Agonist	0
Buprenorphine	P. agonist	0	0
Pentazocine	Antagonist	P. agonist	Agonist
Butorphanol	Antagonist	P. agonist	Agonist
Nalbuphine	Antagonist[a]	P. agonist	Agonist
Naloxone	Antagonist	Antagonist	Antagonist

[a] May have partial agonist activity.
0 = no activity; p. agonist = partial agonist.

effects of opioids are produced by very similar pharmacological mechanisms. Agonists like morphine or fentanyl probably interact with mu opioid receptors in the cord, brain stem, and limbic cortex to produce most of their analgesic effects. In analgesic doses, all of these opioids produce some degree of respiratory depression, which also results from interaction with mu receptors (in the medulla). It stands to reason that reversal of respiratory depression normally entails some reversal of analgesia.

■ Opioid-induced Respiratory Depression

Opioids cause medullary respiratory centers to become less sensitive to CO_2, even when there is no obvious change in respiratory rate. As administered doses increase, breathing slows and becomes deeper; eventually it may become irregular or even Cheyne-Stokes in nature. High doses of morphine or fentanyl can even cause apnea in a patient who is still awake and responsive.

Equianalgesic doses of all opioids produce equivalent degrees of respiratory depression. There is no convincing evidence that any analgesic is more or less dangerous than morphine in this regard. Both fentanyl and alfentanil have been reported to cause "recurrent" ventilatory depression; episodes have been reported in which patients arrived in recovery areas, apparently breathing adequately, only to suffer respiratory arrest some time later. Becker and

Table 3. *Relative Potencies of Various Opioids as Agonist or Antagonists*

Drug	Analgesic Potency (morphine = 1)	Antagonist Potency (nalorphine = 1)
Morphine	1	0
Buprenorphine	25	10
Butorphanol	5	Weak
Nalorphine	1	1
Nalbuphine	0.5	0.25
Pentazocine	0.25	0.02
Naloxone	0	30

associates [4] demonstrated recurrent depression using serial CO_2 response curves.

We believe that this phenomenon is best explained by the relationship between arousal and drug effect. When a person given an opioid falls asleep, breathing is dramatically depressed. As long as such a patient is moved about and told to breathe (as occurs upon admission to the PACU), the patient will continue to do so—until stimulation ceases.

■ Naloxone

Naloxone is the prototypical pure opioid antagonist. It acts at all opioid receptors, but it has greatest affinity for mu receptors. Naloxone can reverse all the effects of opiates, including analgesia, sedation, respiratory depression, pruritis, biliary colic, and urinary retention. Because naloxone has lower affinity at the kappa receptor, larger doses are needed to antagonize drugs like nalbuphine and butorphanol. Buprenorphine is unusual in that it binds to mu receptors with an affinity similar to that of naloxone. Even very large doses of naloxone cannot reliably reverse the effects of this drug [5]. Naloxone probably has no effect on nonnarcotic anesthetics [6], although this remains somewhat controversial.

Naloxone has no opioid agonist properties and does not cause addiction [7]. Its physiological effects in the absence of circulating opioids or opioid dependency are subtle. In man, large doses (4 mg/kg) cause a mild increase in heart rate and systolic blood pressure [8], as well as slowing of electroencephalographic (EEG) alpha-wave activity [9]. Animal studies have also found that naloxone administration reduces food intake, alters sleep patterns, and improves spatial learning. In some disease states, such as septic shock, large doses of naloxone have a pressor effect. The presumed mechanism involves naloxone's antagonism of the effects of elevated endogenous opioid peptides.

Naloxone's competitive antagonism may be overcome by the administration of additional opioid agonist. The duration of action of parenteral naloxone is very short, because it undergoes rapid redistribution from the brain. Reappearance of the narcotic effect is possible, so continued observation is important. Repeated doses of naloxone are sometimes required, even with the short-acting opioid alfentanil [10].

Naloxone can precipitate withdrawal in opioid-dependent individuals. Even patients legitimately using opioids for extended periods may have low levels of physical dependence and therefore may be susceptible to naloxone-induced withdrawal reactions.

In the absence of prior narcotic administration, naloxone does not cause catecholamine release or hemodynamic perturbations [11]. Opioid reversal, however, can have important hemodynamic consequences [12]. Increases in blood pressure, heart rate, and norepinephrine- and epinephrine release can occur. In animals, this response is worsened by hypercarbia but can be seen in normocarbia, even in the absence of a surgical incision [13]. In humans, there have been several case reports of pulmonary edema and sudden death following naloxone administration. Sudden death has occurred in young, previously healthy individuals [14], and pulmonary edema was described in a young man given only 100 μg of naloxone [15]. The etiology of this rare, catastrophic response is not known.

Naloxone can be administered intravenously or intramuscularly. Naloxone is only 1/50th as effective taken orally as parenterally, because a large proportion is metabolized in the first pass through the liver via the portal circulation. This difference between the oral and parenteral effects is cleverly employed to reduce the abuse potential of one oral analgesic (Talwin NX, a mixture of pentazocine and naloxone). The naloxone has insignificant antagonistic effects when the mixture is taken orally, but it blocks the effects of pentazocine if the oral tablet is illegally pulverized and injected.

The postoperative use of naloxone usually involves one of five clinical settings:

1. Reversal of sedation and respiratory depression
2. Diagnosis of postoperative coma
3. Reversal of opioid-induced biliary colic [16]
4. Reversal of side effects from spinal opiates
5. Reversal of neonatal narcotic depression in the delivery suite

Mild overnarcotization is a fairly common problem in the postoperative setting. As stated previously, doses of opioids that seem appropriate when patients are alert (and concentrating on their pain) may be excessive when they fall back to sleep. These patients can usually be roused verbally and encouraged to breathe. They should also be given supplemental oxygen. Small doses of naloxone (0.04 mg in an adult, repeated every 3 min) can be given intravenously, usually with dramatic improvement. In most cases, there is partial reversal of analgesia as well, but this can be minimized by careful dosing. Patients who receive naloxone need continued observation and possibly repeated doses. Also, respiratory compromise is frequently due to a combination of factors, and therapy with naloxone does not eliminate the need to search for and treat conditions like residual paralysis, bronchospasm, and airway edema.

The use of high-dose morphine or fentanyl anesthesia in cardiac surgery typically produces an extended period of postoperative respiratory depression. The intense intraoperative analgesia is desirable, but prolonged intubation and ventilation may be a source of additional morbidity. In one study, naloxone was used to facilitate early extubation after high-dose morphine anesthesia (2 mg/kg) [17], but the side effects appeared to be unacceptable: Many patients had persistent hypercarbia, vomiting, or bladder distension. Pulmonary edema was also described in another individual who received naloxone after a high-dose morphine anesthetic [18]. If early extubation after cardiac surgery is the desired goal, it makes more sense to choose a smaller dose of opioid at the outset.

Naloxone may be used as a diagnostic aid in patients who are unexpectedly slow to awaken from general anesthesia when opioids have been administered. If naloxone fails to work, attention can quickly be directed away from opioids to other causes such as hypoglycemia or intracerebral pathology. On the other hand, there are reported cases of naloxone's causing temporary arousal from nonopioid central nervous system (CNS) depression [19]. Whether this represents reversal of endorphins or some nonspecific analeptic effect has not been determined. We think such use, though not harmful, is unlikely to be beneficial.

Opioids placed in the subarachnoid or epidural space have proven to be an extremely effective means of postoperative pain control. The complications, however, are more disagreeable, and possibly more dangerous, than those of intravenous or intramuscular injections. Severe pruritus is not uncommon after epidural morphine; early and delayed respiratory depression are much less common, but feared.

A theoretical argument has been made for the routine intravenous infusion of naloxone to prevent these side effects. Analgesia after intrathecal opioid placement is due to extremely high local concentrations in the spinal cord; itching and respiratory depression are presumably due to much lower concentrations in the brain stem. The side effects should therefore be more sensitive to naloxone. Unfortunately, the clinical experience with naloxone infusions has been mixed; some studies have found them to be beneficial [20], whereas others have shown that they produce unacceptable reductions in analgesic [21, 22] effect. Urinary retention is another side effect common in males who have received lumbar epidural opioids. This complication is probably due to drug actions in both brain and spinal cord, and Rawal and co-workers [23] have had some success in preventing it with naloxone.

A special case of opioid-induced respiratory depression occurs in neonates born to mothers treated with narcotics during labor. Clark and colleagues [24] treated a group of depressed babies whose mothers had received meperidine and promethazine. Naloxone, in a dose of about 10 μg/kg, was given via umbilical vein catheter, and ventilation was supported as needed. Nonacidotic infants responded dramatically; acidotic infants were slow to reverse and frequently required a second dose. Although the pharmacokinetics of naloxone in the newborn are not known, none of Clark's patients became renarcotized.

■ Naltrexone

Naltrexone is a relatively pure opioid antagonist available only in an oral preparation. Even though anesthesiologists do not normally prescribe this drug, they should understand how to manage a surgical patient who has been on long-term naltrexone maintenance. Such an individual may present a formidable challenge in pain management.

The main clinical use of naltrexone is in the treatment of previously detoxified heroin addicts. In high doses, naltrexone blocks the euphoriant effects of injected heroin and thus helps to prevent relapse. The drug undergoes 95% first-pass metabolism to 6-beta-naltrexol, an active metabolite that probably accounts for most of the drug's activity.

Acute pain in a patient taking naltrexone should be managed, where possible, with regional anesthesia, nonsteroidal anti-inflammatory analgesics, and other nonopioid methods. If opioids are necessary, naltrexone antagonism is competitive and can be overcome with sufficient doses of agonists like morphine or fentanyl. Occasionally, depending upon the time of the last dose and factors which alter gastrointestinal absorption, extraordinarily large doses of narcotics may be needed. The active metabolite has an elimination half-life of 12.9 hours, so clinically important antagonistic effects may be present for 2 to 3 days.

■ Agonist-Antagonist Opioids for Analgesia

As we have mentioned, drugs like nalbuphine, pentazocine, butorphanol, and buprenorphine are used primarily as analgesics and vary widely in their potency as opioid antagonists (see Table 3). The ratio of agonist to antagonist effects is also highly variable. Their potential uses in anesthesia have been reviewed recently [25].

The most important clinical properties of these drugs may be summarized briefly:

1. All of them behave as partial agonists, producing a limited amount of analgesia and a limited degree of toxicity. They are sufficiently active to treat most moderate-to-severe pain, but they may not adequately suppress intraoperative pain. Several of them appear to be effective when given epidurally.
2. The kappa agonists like butorphanol are highly sedating, even in subanalgesic doses. They do not elevate mood like morphine, and they have low abuse potential.
3. Dangerous levels of respiratory depression are very unusual when these drugs are given to otherwise healthy patients. However, apnea may still occur when agonist-antagonists are combined with other CNS depressants.
4. Naloxone can reverse the effects of all agonist-antagonists except buprenorphine (see above).

■ Agonist-Antagonist Opioids for Reversal of Agonists

Because these drugs are opioid antagonists, they can precipitate withdrawal in opioid-dependent patients, and they can reverse the effects of previously administered agonists. Patients who have been treated with large amounts of morphine, meperidine, or oxycodone compounds may develop low levels of physical dependence which can be unmasked by an antagonist. An agonist-antagonist is probably not the best choice to provide analgesia in such an individual.

Nalbuphine is the only agonist-antagonist that has been studied extensively as a reversal agent. The putative advantage of reversal with nalbuphine is the fact that it has intrinsic analgesic properties [26, 27]. This baseline analgesia is postulated to be useful in preventing pain and excitation even though respiratory depression is antagonized. Unfortunately, the available data suggest that reversal with nalbuphine is no more pleasant or safe than reversal with naloxone. Nalbuphine reversal has been reported to produce pain and excitation on emergence, and even pulmonary edema [28, 29].

■ New Horizons

One of the major drawbacks of naloxone is its short duration of action. In response to the need for longer-acting opioid antagonism, nalmefene, a potent, extremely long-acting pure antagonist, has been developed and is currently undergoing clinical trials. Gal and DiFazio [30] found that 2 mg of nalmefene can block the effects of repeated fentanyl injections for over 8 hours. This drug should certainly prevent renarcotization in most cases, and infusions are unlikely to be necessary. It will need to be used with caution, because it can potentially eliminate all analgesia for a very long time.

The mythical potent analgesic devoid of respiratory depression has not been found, but there are some promising avenues for research. Pasternak and associates [31] have described an antagonist called naloxonazine which (unfortunately) reverses analgesia in rats without reversing respiratory depression. Even though the effect is the opposite of that desired, it shows that the mechanisms are somehow separable. This data has been used to hypothesize the existence of mu-1 and mu-2 receptor subtypes. In theory, a specific mu-1 agonist is the drug we want: an opioid analgesic without respiratory depression. Thus far, such an agent has not been found, although the drug meptazinol is a somewhat selective mu-1 agonist which is in clinical use in Great Britain.

■ Final Thoughts

In general, we feel opioid reversal should be used only when clearly necessary. It is almost always possible to titrate opioids to the point at which pain is tolerable but respiration is not compromised. More severe pain requires more opioid, but the pain itself will offset some respiratory depression. This may not be true in patients with other reasons for their respiratory dysfunction. Hypothermia, chronic obstructive pulmonary disease, metabolic derangements, and residual anesthesia may make it more dangerous to give otherwise adequate doses of opioids. The prudent solution in these cases may be to administer opioids as needed while supporting respiration mechanically.

■ References

1. Lasagna L, Beecher HK. The analgesic effectiveness of nalorphine and nalorphine-morphine combinations in man. J Pharmacol Exp Ther 1954;112:356–363
2. Wikler A, Fraser HF, Isbell H. N-allylnormorphine: effects of single doses and precipitation of acute "abstinence syndromes" during addiction to morphine, methadone or heroin in man (postaddicts). J Pharmacol Exp Ther 1953;109:8–20
3. Martin WR, Eades CG, Thompson JA, et al. The effects of morphine- and nalorphine-like drugs in the non-dependent chronic spinal dog. J Pharmacol Exp Ther 1976;197:517–532

4. Becker LD, Poulson BA, Miller RD, et al. Biphasic respiratory depression after fentanyl-droperidol or fentanyl alone used to supplement nitrous oxide anesthesia. Anesthesiology 1976;44:291–296
5. Gal TJ. Naloxone reversal of buprenorphine-induced respiratory depression (abstract). Anesthesiology 1988;69(3A):A818
6. Duncalf D, Nagashima H, Duncalf RM. Naloxone fails to antagonize thiopental anesthesia. Anesth Analg 1978;57:558–562
7. Jasinski DR, Martin WR, Haertzen CA. The human pharmacology and abuse potential of N-allylnoroxymorphone (naloxone). J Pharmacol Exp Ther 1967;157:420–426
8. Cohen MR, Cohen RM, Picker D, et al. Physiological effects of high dose naloxone administration to normal adults. Life Sci 1982;30:2025–2031
9. Volavka J, James B, Reker D, et al. Electroencephalographic and other effects of naloxone in normal men. Life Sci 1979;25:1267–1272
10. Krane BD, Kreutz JM, Johnson DL, Mazuzan JE. Alfentanil and delayed respiratory depression: case studies and review. Anesth Analg 1990;70:557–561
11. Estilo AE, Cottrell JE. Hemodynamic and catecholamine changes after administration of naloxone. Anesth Analg 1983;61:349–353
12. Patschke D, Eberlein HJ, Hees W, et al. Antagonism of morphine with naloxone in dogs: cardiovascular effects with special reference to the coronary circulation. Br J Anaesth 1977;49:525–533
13. Mills CA, Flacke JW, Miller JD, et al. Cardiovascular effects of fentanyl reversal by naloxone at varying arterial carbon dioxide tensions in dogs. Anesth Analg 1988;67:730–736
14. Andree RA. Sudden death following naloxone administration. Anesth Analg 1980; 59:782–784
15. Partridge L, Ward CF. Pulmonary edema following low-dose naloxone administration. Anesthesiology 1986;65:709–710
16. Lang DW, Pilon RN. Naloxone reversal of morphine-induced biliary colic. Anesth Analg 1980;59:619–620
17. Johnstone RE, Jobes DR, Kennell EM, et al. Reversal of morphine anesthesia with naloxone. Anesthesiology 1974;41:361–367
18. Flacke JW, Flacke WE, Williams GD. Acute pulmonary edema following naloxone reversal of high-dose morphine anesthesia. Anesthesiology 1977;47:376–378
19. Finck AD, Salcman M, Balis E. Alleviation of prolonged postoperative central nervous system depression after treatment with naloxone. Anesthesiology 1977;47: 392–394
20. Daly MY, Woods AM, DiFazio CA. Continuous naloxone infusion following epidural morphine: a dose-response study (abstract). Anesthesiology 1988;69:A683
21. Gowan JD, Hurtig JB, Fraser RA, et al. Naloxone infusion after prophylactic epidural morphine: effects on incidence of postoperative side-effects and quality of analgesia. Can J Anaesth 1988;35:143–148
22. Gueneron JP, Ecoffey CL, Carli P, et al. Effect of naloxone infusion on analgesia and respiratory depression after epidural fentanyl. Anesth Analg 1988;67:35–38
23. Rawal N, Mollefors K, Axelsson K, et al. An experimental study of the urodynamic effects of epidural morphine and naloxone reversal. Anesth Analg 1983;62: 641–647
24. Clark RB, Beard AG, Barclay DL. Naloxone in the newborn infant. Anesth Rev 1975;(Dec):9–11
25. Rosow CE. Agonist-antagonist opioids: theory and clinical practice. Can J Anaesth 1989;36:S5–S8
26. Rawal N, Wennhager M. Influence of perioperative nalbuphine and fentanyl on postoperative respiration and analgesia. Acta Anaesthesiol Scand 1990;34:197–202

27. Bailey PL, Clark NJ, Pace NL, et al. Antagonism of postoperative opioid-induced respiratory depression: nalbuphine versus naloxone. Anesth Analg 1987;66:1109–1114
28. Jaffe RS, Moldenhauer CC, Hug CC, et al. Nalbuphine antagonism of fentanyl-induced ventilatory depression: a randomized trial. Anesthesiology 1988;68:254–260
29. Desmarteau JK, Cassot AL. Acute pulmonary edema resulting from nalbuphine reversal of fentanyl-induced respiratory depression. Anesthesiology 1986;65:237
30. Gal TJ, DiFazio CA. Prolonged antagonism of opioid action with intravenous nalmefene. Anesthesiology 1986;64:175–180
31. Pasternak GW, Zhong AZ, Tecott L. Developmental differences between high and low affinity opiate binding sites: their relationship to analgesia and respiratory depression. Life Sci 1980;27:1185–1190

Reversal of Neuromuscular Blockade

Frederic M. Ramsey, MD

Intravenous muscle relaxants have been important adjuvants to general anesthetic agents for nearly half a century. When muscle relaxants are used, it is the responsibility of the anesthesiologist to ensure complete clinical recovery of neuromuscular function prior to withdrawal of respiratory support and airway protective measures at the conclusion of surgery. Despite widespread use of reversal agents and the advent of intermediate-duration muscle relaxants, residual neuromuscular blockade, defined as a train-of-four (T4) value less than 0.70, has been estimated to occur in 21 to 42% of patients in the immediate postoperative period [1–3].

Spontaneous recovery of neuromuscular function takes place through redistribution of relaxant molecules away from the neuromuscular junction and elimination by metabolism or excretion of the parent compound. This process is often enhanced pharmacologically when the rate of spontaneous recovery is too slow to ensure complete clinical recovery by the completion of surgery. This chapter reviews the clinical recovery of neuromuscular function and the pharmacology of antagonism of residual neuromuscular blockade (NMB). The key to ensuring adequate clinical recovery resides in assiduously monitoring neuromuscular function in each patient and in providing an appropriate degree of relaxation through careful titration of muscle relaxants to the desired effect.

■ General Concepts

Neuromuscular Margin of Safety

A margin of safety exists at the neuromuscular junction (NMJ) which requires that approximately 75% of the nicotinic acetylcholine (ACh) receptors be blocked before transmission begins to fail at physiological frequencies of transmission. When 90% or more of the receptors are occupied, transmission is blocked completely even at very low frequencies [4]. Detection of transmission failure using a nerve stimulator is thus possible only within a narrow range (75–90%) of receptor blockade. An important corollary is that, during the recovery phase, a patient who appears clinically recovered may still have a majority of these receptors blocked; although function has been restored, the neuromuscular margin of safety may still be compromised.

Neuromuscular Block vs Receptor Block

It is important to understand the distinction between *percent neuromuscular block* and *percent receptor block,* which are not interchangeable. The relationship between "receptor blockade" and "neuromuscular blockade" is nonlinear. Fifty percent NMB does not occur when 50% of the receptors are blocked.

Percent NMB refers to a quantifiable measurement, such as muscle force of contraction, the integrated electromyograph (EMG) response, or rate of muscle shortening expressed as a percent of the unblocked control response. These are quantified using a force transducer, EMG, or accelerometer, respectively, and can be measured with sophisticated clinical equipment. The single twitch response, which quantifies the degree of NMB, is the gold standard in this regard. For routine clinical purposes, counting the T4 response allows one to estimate the percent NMB over a range between 75 and 100% NMB (see below under Monitoring).

Percent receptor block refers to a pharmacodynamic situation which is not quantifiable in the clinical setting, i.e., there is no way to measure the number or percentage of receptors blocked. The concept is most important in understanding the significance of the neuromuscular margin of safety and the fact that it may be compromised despite recovery of "normal" neuromuscular function. Indeed, even after antagonism of block by pharmacological means, relaxant molecules are still present in the body, and some receptors at the neuromuscular junction remain blocked. Were it not for the large neuromuscular margin of safety, routine postoperative ventilatory support would be needed, making intravenous muscle relaxants impractical.

These concepts are depicted in Figure 1. Neuromuscular function (descending curve) and receptor occupancy (ascending curve) are each plotted as a function of the relative concentration of relaxant. The relationship between relaxant concentration and receptor binding is given by the equation $Y(\%) = X/(X + Kb)$, where X is the relative relaxant concentration and Kb is the dissociation constant for competitive binding [5]. When $X = Kb$, 50% of the receptors will be bound. To make the mathematics simple, Figure 1 has been plotted for $Kb = 1$. At low concentrations of relaxant the receptor occupancy curve is very steep, indicative of efficient receptor binding. Neuromuscular function

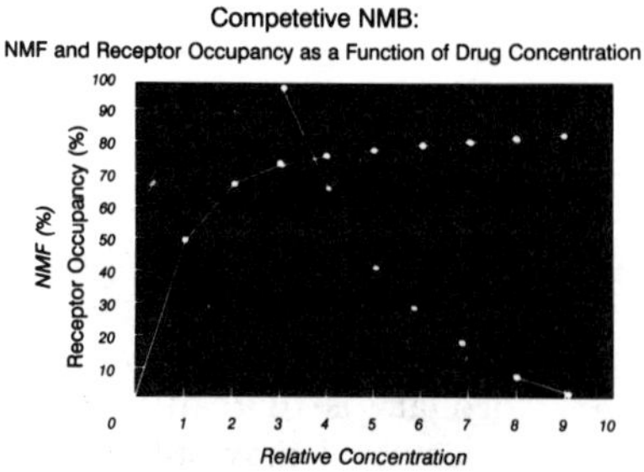

Fig 1. *Neuromuscular function* (descending curve) *and receptor occupancy* (ascending curve), *each plotted as a function of the relative concentration of relaxant. See text for discussion.*

does not start to decline until a relaxant concentration of 3 units ($\dot{X} = 3$) is reached, when 75% of the receptors are bound ($Y = 3/(3 + 1) = 75\%$). A steep decline in neuromuscular function occurs between concentrations of 3 units ($Y = 75\%$) and 9 units ($Y = 90\%$) where the receptor occupancy curve is much flatter (i.e., there is less efficient binding of receptors). Receptor binding increases by only 20% over this concentration range, whereas neuromuscular function declines by 100%! The rightward shift in the neuromuscular function curve reflects the neuromuscular margin of safety described above. These relationships are important considerations as neuromuscular function returns, and they are useful in assessing the adequacy of spontaneous recovery (see below, Is Reversal Always Necessary?).

Mechanism of Reversal

Recovery of neuromuscular function following competitive (nondepolarizing) NMB occurs due to redistribution of relaxant molecules away from nicotinic receptors at the NMJ. This occurs spontaneously as the plasma concentration of relaxant declines during the elimination phase and a concentration gradient develops, driving relaxant molecules from the site of action back into plasma. This washout or redistribution of relaxant can be expedited pharmacologically by enhancing the rate of drug transfer (Fig 2).

The neurotransmitter ACh and nondepolarizing muscle relaxant (NDMR) exist in dynamic equilibrium, competing for the same receptor binding sites at the NMJ. Anything that increases the concentration of ACh or decreases the concentration of NDMR in the synaptic cleft will favor reversal of NMB. Anticholinesterase (antiChE) agents inhibit the enzyme acetylcholinesterase (AChE), thereby extending the half-life of ACh and raising its concentration in the synaptic cleft. By mass action, ACh displaces relaxant from the receptors, and relaxant molecules flow along a concentration gradient back into plasma and are redistributed via plasma washout. AntiChE agents also have prejunctional effects, causing stimulation of motor nerve terminals that leads to enhanced mobilization and release of ACh, further promoting antagonism of NMB [6].

In addition, a reduction in plasma relaxant concentration promotes establishment of a concentration gradient that facilitates transfer of relaxant molecules away from the site of action. If the plasma relaxant concentration is high, the gradient for washout will be low and antagonism may be difficult. This

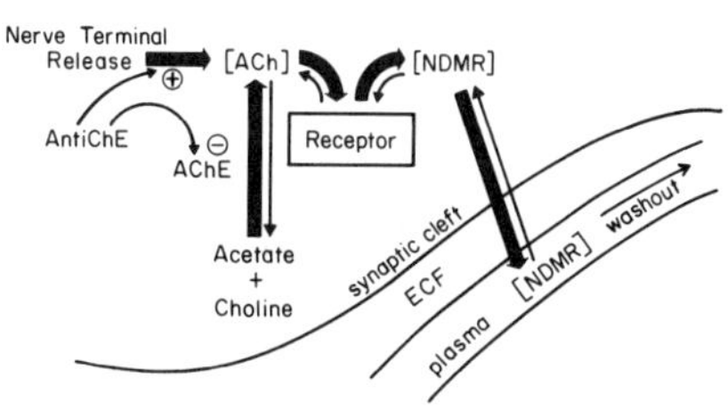

Fig 2. *Pharmacodynamics of anticholinesterase therapy and pharmacokinetics of relaxant redistribution following antagonism of nondepolarizing neuromuscular block. See text for discussion. From Barash PG, Cullen BF, Stoelting RK. Clinical anesthesia. Philadelphia: Lippincott, 1989. Used with permission.*

situation can occur when reversal is attempted soon after administration of relaxant. The longer the time between attempted reversal and the last administration of relaxant, the more successful and prompt reversal is likely to be. During this interval, some degree of spontaneous recovery occurs as the plasma relaxant concentration declines.

The time required to antagonize residual NMB is directly related to the intensity of blockade when reversal is attempted [7]. Although modest degrees of block can be reversed in several minutes, reversal of profound blockade can require 20 to 30 minutes or more. The intermediate-duration relaxants atracurium and vecuronium appear to reverse more readily than longer-acting relaxants [8, 9], which is probably related to their shorter elimination half-lives, rather than to an inherent advantage in reversibility. Thus, recovery time following reversal is a function of the reversal agent used and the spontaneous recovery rate of the relaxant used. A profound block even from atracurium or vecuronium can be very difficult to reverse unless the plasma relaxant concentration has declined to a level that promotes a gradient for effective washout. Thus, recovery time following reversal is a function of two processes: the spontaneous recovery rate of the relaxant used and enhanced recovery promoted pharmacologically with an antiChE agent.

■ Monitoring Neuromuscular Function

Difficulty with reversal is best prevented by avoiding an overdose of relaxant near the end of surgery. Because each patient has a unique dose-response to muscle relaxants, the only way to determine an individual's sensitivity to relaxants is to monitor neuromuscular function throughout the course of anesthesia. Once the initial response has been determined, future bolus doses of relaxant or an appropriate infusion rate can be titrated to the individual's response and surgical requirements. The greatest value of monitoring is derived from titrating muscle relaxation and thus avoiding an overdose near the end of surgery. Monitoring, in other words, should be used to avert reversal problems, not just to confirm recovery.

The most useful clinical parameter to monitor is the T4 response. An approximation of the depth of block is obtained by counting the number of responses elicited by T4 stimulation of the ulnar nerve, as shown in Table 1 [10]. Throughout the course of anesthesia the anesthesiologist should strive to maintain at least one visible twitch to T4 stimulation in order to follow a measurable patient response. Often the response is completely obliterated by the first loading dose of relaxant prior to tracheal intubation; however, once the first response (T1) reappears, subsequent relaxant administration should be adjusted to maintain one or two responses to T4. As a goal, three or four responses to T4 should be visible at the time of antagonism of the block.

Although T4 stimulation is more sensitive than the single twitch response for detecting mild degrees of block, the inability of the clinician to discern significant fade in the fourth response when it is only 40 to 70% of T1 permits mild degrees of residual block to go unnoticed [3]. A new mode of monitoring called double burst stimulation (DBS) consists of two short bursts of tetanic stimulation (50 Hz × 60 ms) separated by 0.75 seconds. The advantage of

Table 1. *Correlation of Train-of-four Count with Depth of Neuromuscular Blockade*

T4 Count	% NMB[a]	Clinical Conditions
0	100	Total peripheral paralysis
1	90–99+	Profound block; abdominal relaxation balanced technique
2	80–90	Deep block; abdominal relaxation inhalation technique
3	75–80	Moderate block; relaxation for peripheral procedures
4 (fade)	5–75	Light block; impairment of respiratory mechanics
4 (no fade)	<5	Clinical recovery of function; decreased margin of safety

[a]% NMB refers to single twitch.
T4 = train-of-four; NMB = neuromuscular blockade.

DBS is that it permits clinical detection of fade between the two responses when the T4 value approaches 70% [11].

■ Pharmacology of Antagonism of Neuromuscular Blockade

Clinically useful antiChE agents include neostigmine, pyridostigmine, and edrophonium. Although neostigmine and pyridostigmine bind covalently with AChE to form an inhibited enzyme-substrate complex, edrophonium binds electrostatically in a reversible fashion with the enzyme [12]. The electrostatic interaction with edrophonium causes enzyme inhibition to be weaker and of shorter duration than that with neostigmine or pyridostigmine. For this reason, edrophonium was not until recently considered a useful reversal agent. In fact, in low doses, edrophonium has a short duration of action and recurarization is possible. If the dose of edrophonium is increased, however, sustained antagonism of NMB from even long-acting relaxants is obtained [13, 14]. Although the pharmacodynamics of edrophonium are distinctly different from those of neostigmine or pyridostigmine, its pharmacokinetic parameters are similar to those of neostigmine and pyridostigmine. Edrophonium, neostigmine, and pyridostigmine have similar elimination half-lives, all of which exceed the elimination half-lives of the long-acting relaxants [15]. Therefore, when an adequate dose of edrophonium has been given, its plasma concentration remains high enough throughout the period of reversal to sustain antagonism of NMB.

Figure 3 helps explain the sustained action of edrophonium when given at high doses (0.5–1.0 mg/kg) to antagonize residual NMB. At low doses (lower curve) the duration of effect is short because the plasma level of edrophonium declines through the therapeutic range during the steep distribution phase rather than the elimination phase of edrophonium. At higher doses of edrophonium (upper curve) the slower fall through the therapeutic range during the beta elimination phase leads to a considerably longer therapeutic effect. As long as the beta half-life of edrophonium is longer than that of the relaxant being antagonized, and as long as it is given in sufficient doses, reversal should be sustained.

AntiChE agents have potent muscarinic effects on the heart (SA and AV node slowing), bronchi (bronchoconstriction and secretion production), and

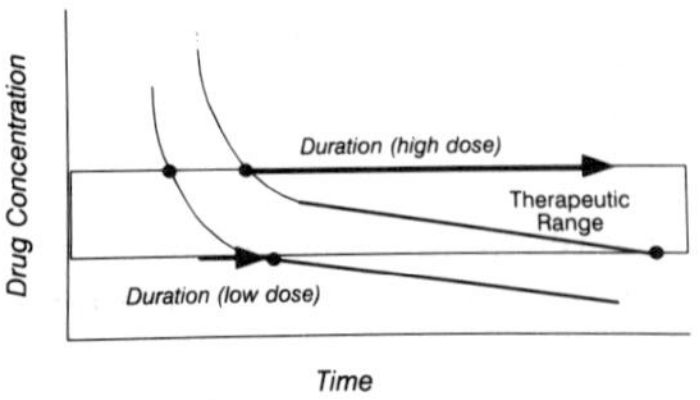

Fig 3. *Effect of reversal agent dosage on duration of therapeutic effect. See text for discussion.*

gastrointestinal tract (peristalsis). Cholinergic side effects can be minimized by either prior or simultaneous administration of an anticholinergic agent. The interplay between cholinergic (antiChE) and anticholinergic effects determines the resultant autonomic response. Glycopyrrolate has a time course (onset and duration) similar to that of neostigmine or pyridostigmine, whereas atropine is better matched with edrophonium. The simultaneous administration of glycopyrrolate with either neostigmine or pyridostigmine leads to minimal net change in heart rate [16], which is particularly important in patients with a history of myocardial ischemia or valvular stenosis. When atropine is administered with edrophonium, it is advised that atropine be given prior to edrophonium to prevent potential bradyarrhythmias caused by edrophonium [17]. Although the vagotonic effect of edrophonium has a very rapid onset, the muscarinic effects of edrophonium are less than those of neostigmine or pyridostigmine. Therefore, less atropine is required with edrophonium than an equipotent dose of glycopyrrolate with neostigmine or pyridostigmine [18].

Guidelines for Antagonism of Neuromuscular Blockade

Recommendations regarding drug doses for antagonism of NMB are shown in Table 2 [19]. The doses of reversal agents should be chosen based on the depth of block at the time of reversal, as determined by the T4 count. Edrophonium is not recommended for antagonism of deep NMB when fewer than two responses to T4 stimulation are visible. Neostigmine is more effective than edrophonium at promoting complete reversal from a profound degree

Table 2. *Dose Recommendations for Antagonism of Neuromuscular Blockade*

Depth of NMB	T4 Count	Neostigmine (μg/kg) (glycopyrrolate)	Pyridostigmine (μg/kg) (glycopyrrolate)	Edrophonium (μg/kg) (atropine)
Shallow	4	25 (5)	100 (5)	500 (10)
Moderate	2–3	50 (10)	200 (10)	1,000 (10)
Deep	0–1	75 (15)	300 (15)	NR

NMB = neuromuscular blockade; T4 = train-of-four; NR = not recommended.
From Barash PG, Cullen BF, Stoelting RK. Clinical anesthesia. Philadelphia: Lippincott, 1989. Used with permission.

of block [20–22]. If the maximum recommended doses are not effective, additional antiChE is unlikely to further augment recovery and may lead to enhanced side effects.

Hazards of Reversal

The greatest hazard of the reversal process is inadequate antagonism of NMB leading to residual weakness and ventilatory impairment. In addition, the astute clinician will be on guard for side effects from an imbalance in the autonomic effects of antiChE and anticholinergic agents. No combination of anticholinergic/antiChE agents, and no sequence of administration, guarantees prevention of side effects. Although the side effects are usually well tolerated without great risk to the average patient, they can have greater importance in patients with preexisting disease states (e.g., bronchospasm in asthmatic patients or tachycardia in patients with myocardial ischemia).

The anesthesiologist should pay careful attention to the electrocardiogram (EKG), because there is a high incidence of dysrhythmias during and following reversal of NMB. Most of these are innocuous, transient, atrial or junctional dysrhythmias; however, ventricular ectopy and high grade heart block, including complete heart block, may also occur. Guidelines for reversal of nondepolarizing NMB are listed in Table 3.

Antagonism of Phase II Block

Till now our discussion has centered on antagonism of a competitive block from nondepolarizing relaxants; recovery from a depolarizing block provided by succinylcholine (SCh) is different. During recovery from a depolarizing block all four responses to T4 stimulation recover together, i.e., there is no fade. If fade to T4 becomes evident following SCh, then the nature of the block has changed and a dual block or phase II block has occurred. Reversal of NMB from SCh is generally not required because spontaneous recovery is so rapid; however, in the setting of phase II block, spontaneous recovery may be prolonged. Neostigmine and edrophonium have been shown to facilitate recovery from an SCh-induced phase II block [23, 24]. The development of phase II block is quite variable. It can occur at relatively low total doses of SCh (2–5 mg/kg) [25], whereas some patients receiving up to 15 mg/kg may not

Table 3. *Guidelines for Reversal of Nondepolarizing Block*

1. Monitor neuromuscular function: T4
2. Maintain a visible response (T1–2/T4)
3. T3–4/T4 visible prior to attempting reversal
4. Drugs and doses dependent on depth of block
5. Allow sufficient time for reversal
6. T4 equal; 50 Hz tetanus sustained
7. Assess clinical criteria

T4 = train-of-four.
From Barash PG, Cullen BF, Stoelting RK. Clinical anesthesia. Philadelphia: Lippincott, 1989. Used with permission.

manifest it [23]. This further highlights the importance of monitoring neuromuscular function in all patients. Guidelines for antagonism of a phase II block are listed in Table 4. Two precautions should be noted:

1. Because antiChE agents also inhibit pseudocholinesterase, a reversal agent should not be administered if SCh is still present in plasma, lest its metabolism be inhibited and the block prolonged. Waiting for a plateau in recovery permits spontaneous recovery from the depolarizing component of the block to occur.
2. Phase II block is not as easily antagonized as a nondepolarizing block. Thus, a profound block with only one or two twitches on T4 stimulation may not be completely antagonizable. It is better to wait until all four responses are visible, albeit with fade, before attempting to antagonize the block.

■ Clinical Assessment of Recovery

It is incorrect to assume that a spontaneously breathing patient who maintains a normal $PaCO_2$ while breathing through an endotracheal tube has achieved sufficient clinical recovery of neuromuscular function to allow extubation. Even the ability to generate a vital capacity of 10 to 15 ml/kg and an inspiratory force of -25 cm H_2O is possible at a time when the ability to maintain an unobstructed airway is compromised [26]. A good clinical test of adequate recovery is the ability of the patient to sustain a head lift for 5 seconds, which correlates with a T4 ratio >0.70 and a negative inspiratory force of -50 cm H_2O. When a patient remains too deeply anesthetized to perform this maneuver on command, a sustained contraction to a tetanic stimulus at 50 Hz for 5 seconds is a suitable alternative. Other movements of the patient, especially movement against gravity (reaching for the tube, etc.) and coordinated respiratory movements in which the abdominal and chest wall excursions act in synchrony provide useful confirmatory clinical signs.

Clinical factors that may retard recovery of neuromuscular function and the ability to antagonize residual NMB are shown in Table 5. A common clinical condition at the completion of surgery, hypothermia is associated with peripheral vasoconstriction and a decrease in blood flow to skeletal muscle. This redistribution of blood flow can impair delivery of reversal agents to the site of action and retard movement of relaxant molecules away from the NMJ.

When antagonism of NMB is ascertained to be incomplete, the safest

Table 4. *Guidelines for Reversal of Phase II Block*

1. Monitor neuromuscular function: T4
2. Spontaneous recovery; plateau phase (fade in T4)
3. >10 min since last succinylcholine
4. Trial of edrophonium (5 mg; with atropine)
5. Additional edrophonium or neostigmine

T4 = train-of-four.

Table 5. *Factors Associated with Difficult Reversal of Neuromuscular Blockade*

1. Relative overdose
2. Renal failure
3. Hepatic failure
4. Acid-base balance
5. Hypothermia
6. Electrolytes
7. Drug interactions

course is to maintain a protected airway with an endotracheal tube and provide ventilatory assistance if necessary until the patient manifests complete clinical recovery. It is important to remember that, even when the listed criteria have been met, receptors are still blocked at the NMJ and the neuromuscular margin of safety is compromised. In addition to providing supportive therapy, the anesthesiologist should seek other contributing factors that might be treated or help explain the dilemma.

Is Reversal Always Necessary?

With the advent of intermediate-duration nondepolarizing muscle relaxants, many clinicians have questioned the routine use of antagonists to reverse residual NMB. One argument in favor of antagonism is that, even when spontaneous recovery appears complete, the neuromuscular margin of safety may still be compromised. The margin of safety can be improved by administering an antagonist, thus freeing subclinical receptors that are blocked by relaxant molecules present in low concentration at the NMJ. As noted earlier (see Fig 1), although neuromuscular function measured by the single twitch response may be "normal" (100%), as many as 75% of the receptors may still be blocked.

This concept is useful in assessing the rate and adequacy of spontaneous recovery. Figure 4 reproduces the receptor occupancy curve from Figure 1 and highlights several key points in the recovery process. The horizontal band from 75 to 90% receptor binding occurs between relative relaxant concentrations of 3 and 9 units, respectively—i.e., it requires approximately three times the concentration of relaxant to produce complete NMB as it does to produce minimal block. It is possible to detect both these extremes with T4 monitoring. If the interval during spontaneous recovery from when T1 first becomes visible (beginning of recovery) until T4 appears to no longer show fade (recovery "complete") is defined as the clinical half-life of the drug, then the true (pharmacokinetic) half-life, represented by the interval between concentrations 4 and 8, will always be overestimated by the clinical half-life. Therefore, although time is not shown directly in Figure 4, it is possible to clinically estimate the half-life of any relaxant in each patient monitored. Once clinical recovery is "complete" by monitoring criteria, then allowing an additional period of time to permit subclinical receptors to become free may possibly establish a sufficient margin of safety that reversal may be unnecessary.

Should the anesthesiologist elect not to administer an antagonist, this au-

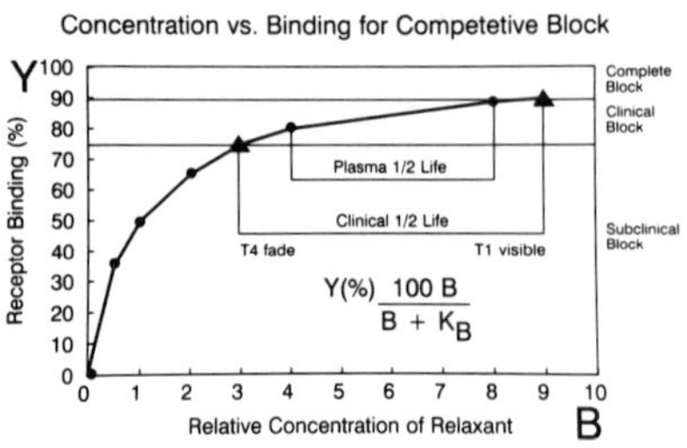

Fig 4. *Estimation of true plasma half-life by using T4 monitoring to measure clinical half-life during spontaneous recovery from nondepolarizing neuromuscular block. See text for discussion.*

thor suggests that at least one additional clinical half-life beyond the appearance of "complete" recovery by T4 criteria be awaited before extubating the patient's trachea. Thus, if "complete" spontaneous recovery of NMB by T4 criteria following atracurium takes 25 minutes, an additional 25 minutes should pass before electing not to administer reversal agents. If an additional clinical half-life has not transpired following "complete" recovery, and tracheal extubation is desired, this author recommends giving one of the drug combinations shown in Table 2 to antagonize shallow NMB. The patient should meet all clinical criteria of recovery in addition to the monitoring criteria above. Other factors, including patient physical status, the risk of side effects from reversal agents, and the effects of the surgical procedure on the patient's ventilatory reserve should be considered as well. If the clinician elects not to administer reversal agents, full recovery of neuromuscular function should be documented in the patient's anesthetic record.

■ Summary

Intelligent, safe use of muscle relaxants dictates that the clinician monitor neuromuscular function in all patients to determine each patient's sensitivity to relaxants. Restoration of muscle strength is a function of pharmacological antagonism of residual NMB, spontaneous recovery as the concentration of relaxant declines at the neuromuscular junction, or both. In addition to recovery of "normal function" by clinical and monitoring criteria, some margin of safety of neuromuscular function must be restored. As new relaxants with very short half-lives become available, it is likely that steady states of relaxation will increasingly be maintained with continuous infusions. In this setting, the rapid rate of spontaneous recovery of both clinical neuromuscular function and an adequate margin of safety may vastly reduce the need for pharmacological antagonism.

■ References

1. Bevan DR, Smith CE, Donati F. Postoperative neuromuscular blockade: a comparison between atracurium, vecuronium, and pancuronium. Anesthesiology 1988;69: 272–276

2. Beemer GH, Rozental P. Postoperative neuromuscular function. Anaesth Intensive Care 1986;14:41–45
3. Viby-Mogensen J, Jorgensen BC, Ording H. Residual curarization in the recovery room. Anesthesiology 1979;50:539–541
4. Waud BE, Waud DR. The relation between the response to "train-of-four" stimulation and receptor occlusion during competitive neuromuscular block. Anesthesiology 1972;37:413
5. Waud BE. Neuromuscular blocking agents. Current problems in anesthesia and critical care medicine, vol 1. Chicago: Year Book, 1977
6. Riker WF. Prejunctional effects of neuromuscular and facilitatory drugs. In Katz RL, ed. Muscle relaxants. Amsterdam: Excerpta Medica, 1975:59
7. Katz RL. Clinical neuromuscular pharmacology of pancuronium. Anesthesiology 1971;34:550
8. Stirt JA, Murray AL, Katz RL, et al. Atracurium during halothane anesthesia in humans. Anesth Analg 1983;62:207–210
9. Williams A, Gyasi H, Melloni C, Bevan DR. Clinical experience with ORG NC45 (Norcuron) as the sole muscle relaxant. Can Anaesth Soc J 1982;29:567–572
10. Lee C. Train-of-four quantitation of competitive neuromuscular block. Anesth Analg 1975;54:649
11. Olsen NV, Drenck NE, Engbaek J, et al. Clinical evaluation of residual curarization: a comparison of train-of-four stimulation and double burst stimulation. Acta Anaesthesiol Scand 1987;31(suppl 86)
12. Taylor P. Anticholinesterase agents. In: Gilman AG, Goodman LS, Rall TW, et al, eds. The pharmacologic basis of therapeutics. 7th ed. New York: Macmillan, 1985:110
13. Kopman AF. Edrophonium antagonism of pancuronium-induced neuromuscular blockade. Anesthesiology 1979;51:139
14. Bevan DR. Reversal of pancuronium with edrophonium. Anaesthesia 1979;34:614
15. Morris R, Cronnelly R, Miller RD, et al. Pharmacokinetics of edrophonium and neostigmine when antagonizing *d*-tubocurarine neuromuscular blockade in man. Anesthesiology 1981;54:399
16. Salem MG, Richardson JC, Meadows GA, et al. Comparison between glycopyrrolate and atropine in a mixture with neostigmine for reversal of neuromuscular blockade. Br J Anaesth 1985;57:184
17. Urquhart ML, Ramsey FM, Royster RL, et al. Heart rate and rhythm following an edrophonium/atropine mixture for antagonism of neuromuscular blockade during fentanyl/N_2O/O_2 or isoflurane/N_2O/O_2 anesthesia. Anesthesiology 1987;67:561
18. Cronnelly R, Morris RB, Miller RD. Edrophonium: duration of action and atropine requirement in humans during halothane anesthesia. Anesthesiology 1982;57:261
19. Lebowitz PW, Ramsey FM. Muscle relaxants. In: Barash PG, Cullen BF, Stoelting RK, eds. Clinical anesthesia. Philadelphia: Lippincott, 1989:339–370
20. Kopman AF. Recovery times following edrophonium and neostigmine reversal of pancuronium, atracurium, and vecuronium steady-state infusions. Anesthesiology 1986;65:572
21. Rupp SM, McChristian JW, Miller RD, et al. Neostigmine and edrophonium antagonism of varying intensity neuromuscular blockade induced by atracurium, pancuronium, or vecuronium. Anesthesiology 1986;64:711
22. Caldwell JE, Robertson EN, Baird WL. Antagonism of vecuronium and atracurium: comparison of neostigmine and edrophonium administered at 5% twitch height recovery. Br J Anaesth 1987;59:478
23. Ramsey FM, Lebowitz PW, Savarese JJ, et al. Clinical characteristics of long-term succinylcholine neuromuscular blockade during balanced anesthesia. Anesth Analg 1980;59:110

24. Donati F, Bevan DR. Antagonism of phase II succinylcholine block by neostigmine. Anesth Analg 1985;64:773
25. Lee C. Dose relationships of phase II, tachyphylaxis, and train-of-four fade in suxamethonium-induced dual neuromuscular block. Br J Anaesth 1975;47:841
26. Pavlin EG, Holle RH, Schoene RB. Recovery of airway protection compared with ventilation in humans after paralysis with curare. Anesthesiology 1989;70:381

Recovery of Psychological Function after Anesthesia

Lalage D. Sanders, PhD

There has been growing interest within anesthesiology in the psychometric assessment of recovery after drugs, often involving the use of tools more commonly found in the hands of the psychologist. This situation has potential benefit for both anesthesiology and psychology—and has drawn attention to the fact that the two disciplines have a common focus: awareness. The psychologist attempts to understand awareness, whereas the anesthesiologist seeks to control it (i.e., the patient's awareness of a procedure, of pain, or of both).

Two factors account for this relatively new state of affairs. Until recently it was generally accepted that patients undergoing surgery with general anesthesia required several days' bed rest postoperatively. This is no longer true. Patients are now encouraged to become active as soon after surgery as they can. New, rapidly acting anesthetic agents that are also metabolized speedily enable patients to be ambulatory sometimes within hours following surgery. Indeed, increasing numbers of procedures are performed on an out-patient basis. With use of the new anesthetic agents, it has been noted that patients recover more quickly both in the early postoperative phase and in the medium term. Differences that are apparent in the recovery room do not require formal testing, but over the longer term such differences may be more subtle and can be demonstrated only with careful psychometric testing.

The second development that has fueled interest in psychometric testing is the increasing number of procedures now done on a same-day basis. The procedures themselves are unlikely to leave the patient seriously impaired. It is the drugs used for the procedure—the anesthetics, analgesics, or sedatives—that affect the patient. The rate at which the patient recovers from these drugs thus determines the time of release. Because these drugs are centrally acting, they do cause impairment of both cognitive and psychomotor functions, hindering overall performance. It has therefore become important to find reliable means of assessing recovery of psychomotor and cognitive function.

The Meaning of *Recovery*

A prior step, however, is defining the term *recovery*. The *Oxford English Dictionary* defines *recover* as "to cease to feel the effects of." This has a common-sense ring and is certainly satisfactory for lay usage, but it is inadequate for clinical purposes. Ceasing "to feel" has connotations of a subjective evaluation, which may not be a reliable indicator. Recovery following the administration of sedative, analgesic, or anesthetic drugs is more appropriately defined as "return to the pre-drug condition."

There are generally considered to be three phases of postoperative recovery. In the first phase there is recovery of vital reflexes such that the patient may safely be left without skilled attendants; this is the province of the recovery room, with its specially trained personnel. The second phase refers to the time when the patient has regained sufficient cognitive and physical competence to be allowed to return home. Following major surgery this second phase may be dominated by disturbances due to the surgery itself and to consequent metabolic derangement: impairment caused by the anesthetic is less often a major factor in this phase. The third and final phase of recovery is when the patient has returned to the pre-drug state and has no residual functional impairment. At this point the patient is assumed fit to resume normal duties and activities.

In many articles concerning recovery of performance, particularly with reference to day-cases, one finds the term *street fit.* By this is meant a state of recovery that differs from full recovery, yet in which the patient is considered fit to be released onto the streets. But this concept does not withstand close scrutiny: if a "street fit" patient is not fully recovered, just what type of residual impairment can be permitted such that he or she will be safe on the street? Is it reasonable to release someone whose decision-making process is impaired? What happens when a patient has to decide when to cross the road? What about perception? Will a patient be able to perceive the speed of oncoming vehicles correctly? And memory? Will patients remember their instructions concerning what they may or may not safely attempt? And their reasoning? If their reasoning is faulty, can they be trusted to take the right route home?

It can be argued that an individual is "street fit" only when fully recovered; thus the term loses meaning. A more appropriate term for this intermediate stage is *home readiness,* which suggests the patient has recovered sufficiently *to be taken* home by a responsible adult. It may further be argued that if there is no responsible adult to accompany the patient, then release should be delayed.

When patients are assessed prior to release, they are expected to be able to perform certain tasks. It is normally assumed that they should be able to stand unaided, walk a straight line, and be able to state their name and address—assuming they could do these things before surgery. Such indices of recovery are very crude, however. More sophisticated means of assessing functional recovery now exist.

■ Research Studies and Clinical Practice

It has been noted that accurate measurement of recovery is a problem for researchers and clinicians alike. Researchers must evaluate the performance of patients or volunteers in clinical trials: clinicians must evaluate patients in the recovery room, the ward, or the Day Unit.

Assessment of recovery is a complex, time-consuming procedure, however, and can be satisfactorily executed only in a research setting, in which adequate time and facilities are available. It is simply not practicable to conduct such testing in the clinical setting.

In order to do the task properly it is important to take a rigorous, scientific approach. Listed below are several prerequisites to good psychometric research. These points are, of course, in addition to the general rules that apply to all clinical trials. The quality of the literature on this subject varies widely; regrettably, many researchers have not observed these simple rules, and their results must be considered questionable. Ignoring any of four main points may invalidate the results and conclusions of a study.

Learning Curves

It is important to consider that learning may affect performance. Many psychomotor and cognitive tests can be learned, and the results of these tests show substantial practice effects. This need not be a problem if adequate practice is allowed beforehand. Once a subject has reached asymptote, baseline measurements can be taken. Tests in which practice effects are particularly pronounced may be suitable only for volunteer studies that allow sufficient practice time. It is generally impracticable to administer lengthy training sessions for patients, but tests that show a small practice effect may be suitable for patient studies if this is factored into the design of the study.

Baseline Measurements

It is essential to take baseline measurements—an obvious point that is often ignored. There is no sense measuring performance after a drug if there are no data on pre-drug performance. This simple precaution is most frequently ignored when it comes to subjective measures. If we want to know how a particular drug affects heart rate, for example, we would not consider simply giving the drug and then measuring heart rate. However, many studies have been published in which an individual is asked to supply subjective information (using, for example, a symptoms checklist or visual analogues) only *after* the drug is used. This provides only half the required information.

Environment

Psychometric test results can be affected by environmental conditions such as lighting, noise, or distractions. This problem arises chiefly in patient studies; it is easier to control the environment in a laboratory. For this reason it is generally advisable in designing a patient study to select only the most robust tests, which are less likely to be so affected.

There are other practical considerations. A patient's movement may be restricted by the flowing sleeves of a hospital gown or by having an intravenous catheter attached, and individual disabilities such as deafness, poor vision, or unfamiliarity with the language can affect test results.

Asymmetrical Transfer Effects

It is common in psychopharmacological experiments to use a cross-over design [1], by which each volunteer undergoes each treatment—as many as five or six—following a Latin square design in order to balance the sequence. This design is popular because each volunteer acts as his or her own control. Unfortunately, this approach overlooks treatment effects, which can have a serious impact on an individual's responses. This problem can be illustrated by taking the simplest design, namely one active drug and one placebo, as an example. By counterbalancing, half the volunteers experience the active drug first and then the placebo; the remaining half experience the placebo first and then the active drug. These experiences are not comparable, however, in that volunteers entering the second phase will have their expectancies of the trial shaped by their experience of the first phase.

This is not to argue that such a design cannot be used. But when it is employed, it is necessary that the results be analyzed for evidence of a drug/order interaction. It has been argued that, ideally, cross-over designs should be used only when the number of volunteers in each group is large enough to allow for a parallel group analysis [2]. In designing a controlled clinical trial of this type, the investigator must consider the duration of the test battery. The most sensitive tests are those that are fairly lengthy, in which the subject is required to attend to the task for prolonged periods. Quite subtle types of impairment are picked up by detecting lapses in concentration. Furthermore, with a short test it is possible to make a concerted effort to overcome and therefore hide impairment. On the other hand, batteries involving lengthy tests are likely to have a high incidence of withdrawals from the trial. These two conflicting demands must be balanced when a trial is designed. The problem is more easily overcome in volunteer studies. The volunteer expects to perform the test selected by the experimenter and, once recruited, rarely withdraws. Lengthy trials with patients present more of a problem, as many patients are reluctant to perform arduous, potentially boring tasks.

■ From Research to Practice

With assessment methods restricted to research programs, the next step is how to present new information to the clinician in such a way that full advantage can be taken of it.

One problem, for example, is the high degree of individual variation in both performance and recovery, a problem that besets this entire area of research [3]. This variation is actually fourfold: people vary not only in their response to a test, both initially and over time, but also in their response to the drug and to the rate at which they recover from it. Standard practice in journal articles has been to publish mean recovery times, or mean performance scores. But if decisions are made about the progress of patients on the strength of those means, then by definition some patients will not have attained that level. For this reason it is very important to consider the range of the results, using mean recovery time only as a guide for overall patient management. In individual cases, it is essential to realize that residual functional impairment may well last longer than indicated by the mean. It can be argued, indeed, that recovery-assessment research should undergo a shift in emphasis.

For example, although the thrust of such studies in the past has been to *assess recovery*, it may well be that a better approach would be *detecting impair-*

ment. To illustrate the practical implications of this change in emphasis, let us hypothesize, for example, the use of intravenous benzodiazepines as sedation for investigative out-patient procedures such as endoscopy. Analysis of mean scores to assess recovery using a standardized psychomotor test may reveal no significant impairment in a group of patients by, say, 2 hours following drug administration. But it cannot be concluded that no patient will be suffering any residual impairment at this point. If the aim of the study is to detect impairment, however, then the results may reveal a small percentage of patients to be experiencing deficits at the 2-hour point.

The implications of the different emphasis are clear. In the former case, patients would be released at 2 hours, without warning of possible residual impairment. In the latter case, some patients might not be released at 2 hours, and those who were might be warned that they could still experience some functional deficit.

Further research is needed to better assess impairment. One limitation to such research, however, is that young, healthy volunteers may recover more quickly than a standard patient group.

Another issue affecting the manner in which such research is applied in the clinical setting is the frequent stark difference between the progress through the hospital system of patients involved in a research program and that of routine clinical patients. For example, a recent study to examine the effects of a benzodiazepine reversal agent on levels of sedation as experienced by out-patient gastroscopy patients included 3 post-drug assessments [4]. All patients were given benzodiazepine intravenously for sedation just prior to the gastroscopy; trial patients were given the reversal agent when the endoscopic procedure was completed. Patients were invited to participate in the trial only if they were not taking any concurrent drug therapy that might interact with the trial drug, and if they were not suffering a disability that would interfere with completion of the test. Consequently, not all patients on a given list were included in the trial.

By the time of the final assessment 3 hours after the administration of the reversal agent, the only patients still in the unit were those involved in the trial. All non-trial patients had gone home. So it was decided to monitor all non-trial patients undergoing gastroscopy over a 5-week period, a total of 85 patients (unpublished observations). Of these, 30 patients left within 1 hour of receiving *the sedative,* and another 40 left between 1 and 2 hours. Only 15 patients remained in the unit for more than 2 hours.

These results were surprising. The phenomenon was more pronounced with patients given diazepam than with those given midazolam, although the doses were comparable, which suggests that midazolam gives a more profound subjective experience of sedation. For routine endoscopy cases intravenous benzodiazepines are administered by the endoscopist, and there is no anesthesiologist present. Patient release is determined by patients themselves announcing their departure to the nursing staff. It must be concluded that the patients themselves felt well enough to leave.

The promptness with which these patients were leaving the hospital, however, means they were still suffering residual effects while traveling home. Such early departure has not been advocated in publications concerning recovery, and this emphasizes the distance between research protocol and clinical practice. Though it is unlikely any research publication could stem this untimely departure of patients, researchers could at least provide the clinician with the appropriate information so that patients could be forewarned of the risks they run in leaving so early.

■ Functional Impairment

Once it is accepted that the major task for research is to detect impairment, it then becomes necessary to specify more accurately the functions likely to be affected. What type of impairment can be expected? In which functions will it be located?

One simple means of categorizing these functions is to distinguish between psychomotor and cognitive functions. However, these categories are not discrete. Psychomotor functions include those having a large motor component, but normally having a cognitive component as well. For example, although walking is obviously a motor process, it also requires that the individual perceive the environment and also understand the instruction to walk.

It is slightly easier to distinguish tasks that are largely cognitive, such as those that require reasoning, thinking, comprehension, perception, and memory. But even most of these functions have a motor element as well. For example, a simple question-and-answer technique requires some motor coordination for the answer to be enunciated.

It is therefore more useful to visualize these functions as a continuum ranging from the simplest, largely motor, responses to the higher, more complex intellectual functions (Fig 1).

All these functions can be impaired by psychoactive drugs, but that does not mean that a given drug affects all the functions. Nor does it mean functions that are impaired will be equally affected. Impairment is likely to be differentially spread across these functions—meaning that the selection of the test battery must be very thorough. If the wrong functions are chosen for testing, it may well be that no sign of impairment will be discovered when in reality other functions are affected. In the first instance assessment of impairment must be completed with a very comprehensive test battery on volunteers. Once the type of impairment associated with a particular drug has been identified by detailed volunteer trials, it is then possible to measure such impairment in studies involving patients using a more restricted and precise battery of validated, objective measures.

■ Objective Measures

The list of objective tests given below is by no means exhaustive; the number of tests that can and have been used to measure impairment is large [5].

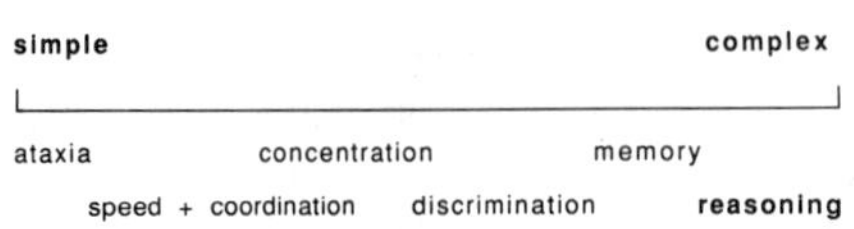

Fig 1. *Continuum of functions.*

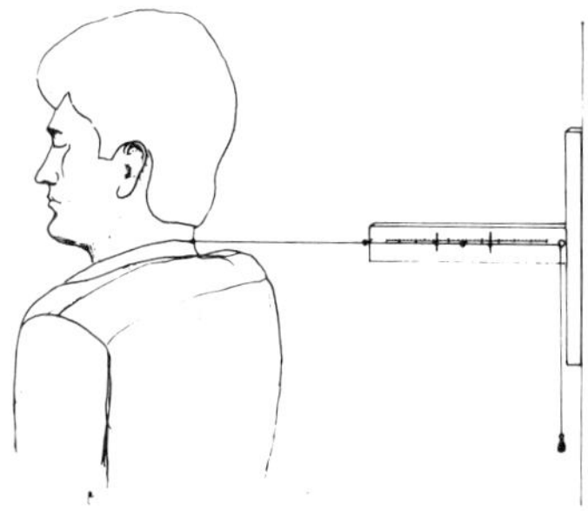

Fig 2. *Simple mechanical measurement of body sway.*

However, the tests presented here are those that are most commonly used and are consistent with the continuum formulation referred to above.

Steadiness

The simplest tests of all are those designed to measure a person's steadiness. It has been known since the seventeenth century that there is movement in the human body at rest. But only relatively recently have the means existed to quantify such movement. Romberg, a nineteenth-century German neurologist, first identified a connection between certain pathological states and the inability to balance with one's eyes closed and feet together, now known as Romberg's sign (or Rombergism). Several factors affect steadiness, such as age, weight, height, and anxiety. But steadiness has been shown repeatedly to be affected by psychoactive drugs [6].

There are many measures of steadiness. The simplest one requires that the individual stand unsupported with his or her eyes closed for a short period of time, without staggering. This is a very crude test, as there are only two possible outcomes: success or failure. There are also mechanical measures designed to gauge the amount of body sway (Fig 2). It is not necessary for these to be complex to be effective. At the far extreme there are highly complex electronic measuring devices that detect small movements of pressure using a balance platform. Video, radar, and ultrasound have also been employed for this purpose.

Speed and Coordination

The next stage of our continuum involves speed and coordination (i.e., actions that are predominantly motor but require a degree of cognitive skill). Tests designed to assess speed and coordination are slightly more complicated. The most frequently used test in this group is reaction time, of which there are many versions. Reaction time is the time taken by an individual to respond to a stimulus. The stimulus may be visual or auditory; there may be a single stimulus or a number of stimuli requiring different responses.

In the same group are the postbox and the peg-board tests. The postbox test is a child's toy, a cylinder in which there are oddly shaped holes; a series of objects must be correctly matched to the holes and inserted. The peg-board task involves moving a number of wooden pins from one side of a piece of peg-board to the other. The scores from these tasks are derived from performance over time, either by recording the length of time required to complete the task or by measuring performance achieved in a specific length of time.

There are also pen-and-paper tests, like the aiming task, which assess the same type of function. In the aiming task a person is given a sheet of paper with a series of small circles connected by lines and is asked to put a dot in the center of as many of these as possible within a given time, working systematically along the lines. The score conventionally includes the total number of dotted circles and an error score for those missed.

Concentration

Tasks that test concentration include the *P* deletion test, in which every letter *P* must be deleted from a section of close-typed upper-case letters. There are variations on this theme, several of which are computerized, and all include an element of vigilance. Another task designed to assess concentration is the digit symbol substitution task, in which a code sheet gives symbols for digits 1 to 9, and the task is to code a series of random numbers within a certain length of time. This task has repeatedly proven sensitive to the effects of the benzodiazepines [7]. These all have a strong motor component as well.

Discrimination

Tests that are largely cognitive and in which the motor component is secondary include, for example, the Critical Flicker Fusion Threshold. The task requires watching a light that is flickering at an increasing rate of frequency and indicating when a flicker can no longer be detected, that is, when the light appears to be on constantly. The test is also administered in reverse, the task being to indicate when a flicker is first detected. This is a popular test which has shown sensitivity to drug effects [7].

There is also a relatively new test called Inspection Time in which the subject is presented with a stimulus for a maximum of 200 milliseconds and must discriminate some feature of that stimulus. There are auditory and visual versions of this test, which claims to measure the speed of cerebral processing and appears to have an element of "real world" validity.

Memory

Memory is a prime indicator of normal function, but many studies inadequately assess amnesia, and inappropriate methodologies are employed that fail to detect existing deficits. It is not enough merely to attempt to examine the patient's recall of the procedure. Nor is it reasonable to show a patient a stimulus object during a procedure with a later requirement to recall that object. It has been shown that salient objects will be recalled more easily than neutral ones, but it is not always possible to assess whether an object is salient to a particular individual. It is more relevant and successful to use reliable tests that measure current memory storage. One test often used to test this function

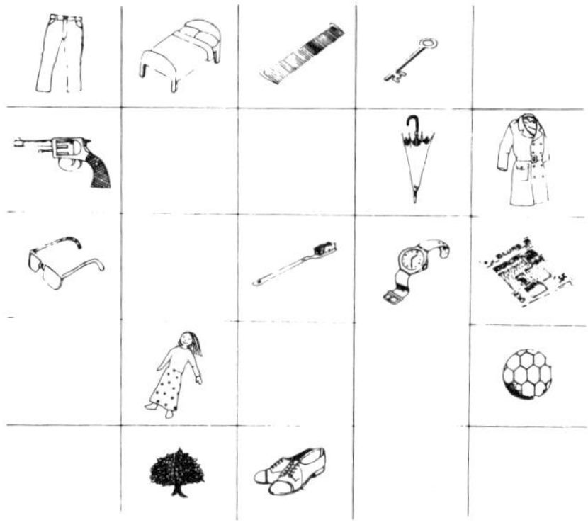

Fig 3. *Object Recall Test Card. An objective measure for short-term spatial recall.*

is the digit span test, which retains an unwarranted popularity although it has seldom proven a sensitive measure of drug-induced amnesia.

Other tests are more effective yet just as simple. There are visual tasks in which a subject is asked to examine briefly a card depicting a number of familiar objects (Fig 3); the card is then removed, and the subject must recall as many of the objects as possible. This test measures short-term spatial recall. There are auditory versions of similar tasks, such as the paired associate test. In this case the subject is read a list of pairs of words and then tested for ability to remember the list. Different versions of this test can be used to assess recognition or recall. In the recall test, the first word from each pair is given and the subject must supply the second. And in the recognition test, the first word from each pair is given followed by four other words, one of which was in the original pair. The task is to identify the correct word. This measure of short-term verbal recognition is more sensitive than the recall task.

Reasoning

At the far end of the continuum is reasoning. Reasoning tests are often quite difficult and are not suitable for all groups of subjects. However, if the subject group can master this type of task, then it can be a sensitive measure of drug effects.

It is important to remember that the functions on our continuum are not uniformly affected by drugs and that full recovery of one function does not imply full recovery of others.

■ Subjective Measures

In addition to objective tests there are also subjective measures, employed to assess the subjective experience. This is arguably a derivation of the oldest form of recovery testing, namely, "And tell me Mrs. Smith, how are you feeling today?" These tests take many different forms; there are check-lists, rating scales, visual analogue scales (VAS), and a variety of validated questionnaires.

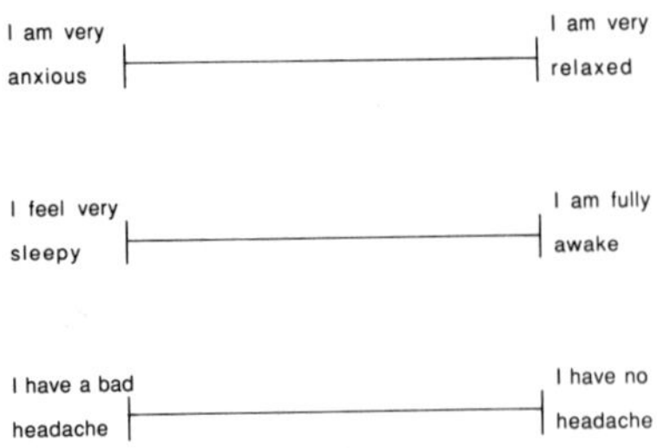

Fig 4. *Visual analogue scales.*

In most instances check-lists, rating scales, and questionnaires are relatively easy to administer, but they lack the flexibility of the VAS. These appear to be the most effective and yet potentially the most troublesome of all subjective recovery tests [8]. As with all other means of assessing impairment, it is necessary to take baseline measurements first, for three reasons. First, as with the objective tests it is impossible to measure drug effects without a pre-drug assessment. Second, there tends to be wide individual variation in these measures; a baseline measure permits analysis that takes individual responses into account. Last, for these tests to provide valid and reliable data, the subject must understand how to complete them. Many people find these tests quite strange at first, so it is necessary to make sure the task is fully explained prior to administration of the drug; afterward, comprehension is likely to be impaired.

If a VAS is used, it is advisable to ensure that the accompanying legend elicits the optimum response. A common problem with this type of test is referred to as the "acquiescent response set," in which the subject tends to answer a series of questions with similar responses, without giving due consideration to each in turn. This effect can be reduced by using meaningful labels on both ends of each analogue (Fig 4).

It is thus possible with careful research to gather data on the subjective response to drugs. However, it is unwise to rely on a subjective measure to judge recovery. As was noted in the survey of gastroscopy patients [5], a substantial number left the Day Unit at a time when research suggests they were still impaired by drugs. In another study patients were asked to indicate by means of a visual analogue whether they felt recovered enough to leave the Day Unit [4]. Even as soon as 40 minutes after receiving the sedative, patients were scoring as "ready to leave."

This raises two important issues with subjective measures of recovery. First, with drugs, as with alcohol, we do not always recognize our own impairment. The drug, in impairing cognitive function, affects one's judgment and self-assessment of performance. Second, other factors influence patients' response to that question—not least the desire to leave the Unit and return home.

■ Conclusion

Recovery after anesthetic and sedative drugs is not adequately understood, and much of the research published in this area falls short of what is needed.

Developing reliable measures of recovery is impossible in the clinical setting, so the task falls to research. There is still no single, simple test of functional impairment, and a comprehensive test battery must be employed to examine all the functions that may be impaired by centrally acting drugs.

The first task for those examining recovery of psychological function must be to learn how to determine when a patient is fit to be released if accompanied. It is also important, however, to be able to assess the duration of residual impairment and the late recovery phase, in order to accurately instruct patients concerning return to normal duties—including driving, for example. This can be achieved only within a rigorous scientific framework, and must produce a data-bank of information on which the clinician can draw. To do this thoroughly, researchers must be aware of the continuum of functions that are differentially affected by different drugs.

■ References

1. Millar K. Some chronic problems in the methodology of performance tasks applied in clinical settings. In: Broughton R, Ogilvie RD, eds. Sleep, arousal and performance. Cambridge, MA: Birkhäuser (in press)
2. Armitage P, Hills M. The two-period crossover trial. The Statistician 1982;31: 119–131
3. Herbert M. Assessment of performance in studies of anaesthetic agents. Br J Anaesth 1978;50:33–38
4. Sanders LD, Piggott SE, Isaac PA, et al. Reversal of benzodiazepine sedation with antagonist flumazenil. Br J Anaesth (in press)
5. Hindmarch I. Psychomotor function and psychoactive drugs. Br J Clin Pharmacol 1980;10:189–209
6. McClelland GR. Body sway. In: Klepper ID, Sanders LD, Rosen M, eds. Ambulatory anaesthesia and sedation. Oxford: Blackwell (in press)
7. Wittenborn JR. Behavioural toxicity in normal humans as a model for assessing behavioural toxicity in patients. In: Lipton MA, DiMascio A, Killam KF, eds. Psychopharmacology: a generation of progress. New York: Raven, 1978:791–796
8. Robinson JO, Sanders LD. Subjective assessment. In: Klepper ID, Sanders LD, Rosen M, eds. Ambulatory anaesthesia and sedation. Oxford: Blackwell (in press)

Index

Future Issues

Volume 29, 1991

Summer (No. 3)

ORGAN TRANSPLANTATION

Fall (No. 4)

CONSTANT INFUSIONS FOR MAINTAINING ANESTHESIA